Pastoralism and Climate Change in East Africa

Edited by

P.Z. Yanda and C.G. Mung'ong'o

MKUKI NA NYOTA

DAR–ES–SALAAM

PUBLISHED BY
Mkuki na Nyota Publishers Ltd
P. O. Box 4246
Dar es Salaam, Tanzania
www.mkukinanyota.com

For:
Centre for Climate Change Studies (cccs)
University of Dar es Salaam
P.O. Box 33453
Dar Es Salaam
www.cccs.udsm.ac.tz

ISBN 978-9987-75-392-5

Visit www.mkukinanyota.com to read more about and to purchase any of Mkuki na Nyota books.
You will also find featured authors, interviews and news about other publisher/author events.
Sign up for our e-newsletters for updates on new releases and other announcements.

Distributed worldwide outside Africa by African Books Collective.
www.africanbookscollective.com

Contents

FOREWORD. VII

ABOUT THE AUTHORS IX

PART I:
INTRODUCTION 1

CHAPTER 1:
Pastoralism for a Sustainable Future in
East Africa   An Introduction C.G. Mung'ong'o and P.Z. Yanda. . . . 3

CHAPTER 2:
A Photographic Introduction to Pastoralism
in Tanzania by Vedasto Msungu, photographs
taken in 2016 in northern Tanzania 9

CHAPTER 3:
Understanding the Pastoral Production System
of East Africa Abdelaziz Gaiballah and Nancy
Ibrahim Abdalla 33

CHAPTER 4:
The Impacts of Climate Change on the Pastoral
System in East Africa by Nancy Ibrahim Abdalla
and Abdelaziz Gaiballah. 49

PART II:

Vulnerability and Responses to Climate
Change Shocks and Stresses in Pastoral Communities. 67

CHAPTER 5:

Dimensions of Vulnerability to Climate
Change and Variability in Pastoral Production
Systems of East Africa Anthony Egeru, by Carloyne
Nandozi, Angella Kyobutungi and J.R.S. Tabuti. 69

CHAPTER 6:

Role of Culture and Gender in Shaping
Rangeland Management and Adaptation
by John R.S. Tabuti, Anthony Egeru and Emmy Jakisa. 85

PART III:

Pastoral Knowledge Systems
and Resilience to Climate Change .105

CHAPTER 7:

The Technical and Social Knowledge Systems of
Pastoral Mobility in East Africa by Emma T. Liwenga 107

CHAPTER 8:

Nomadic Pastoralism and Natural Resource Use Conflicts in
East Africa by Paul Onyango 123

CHAPTER 9:

Policies, Legal Frameworks and Institutions Governing
Pastoral Systems in East Africa by Adolphine Kateka 157

PART IV:

Analysis of the Contribution of Pastoralism
to the National Economies of East Africa 199

CHAPTER 10:

Pastoralism in East Africa: Economic Trends, Patterns,
Opportunities and Challenges by Edmund Mabhuye 201

CHAPTER 11:

A Tale of Three Villages: A Study of Social Stratification
in Three Pastoral Communities in Simanjiro District,
Northern Tanzania by C.G. Mung'ong'o 227

PART V:

Synthesis and Conclusions 249

CHAPTER 12:

Sustainable Pastoralism in East Africa: A Pipe Dream or
Future Reality? by C.G. Mung'ong'o and P.Z. Yanda 251

LIST OF FIGURES, PLATES AND TABLES

Case Study 10.1:   Simanjiro District

Case Study 10.2:   Simanjiro District

Figure 3.1:   Map of main livestock routes in East Africa

Figure 7.1:   Three pillars of pastoralism

Figure 9.1:   Map of rainfall patterns in Tanzania

Plate 8.1:   A young Borana pastoralist wields a Kalashnikov gun.

Table 3.1:   Profile of the pastoralist production systems, major groups, livestock populations, mobility, rangeland resources and climatic zones

Table 4.1:   Possible climate change impacts on East African rangeland ecosystems

Table 4.2:   A summary of the impacts of climate change on livestock production

Table 4.3:   Adaptation measures taken by pastoralists to deal with climate change impacts

Table 8.1:   An interactive governance framework for understanding conflict in pastoral communities

Table 11.1:   Percent Distribution of Wealth Groups and their Respective Socio-Economic Characteristics in Kitwai A, Namalulu and Terrat Villages, 2009-2015

Foreword

Pastoral livelihoods have become increasingly vulnerable to climate change impacts as a result of prolonged marginalization and harmful external interventions. Traditional pastoralism has proved to be a resilient and unique system of adaptations in a dynamic process of unpredictable climatic variability and continuous human interactions with the natural environment in dryland ecosystems. Pastoral adaptations and climate-induced innovative coping mechanisms are strategically embedded in the indigenous social structures and resource management value systems. However, the case of global climate change, expressed in terms of increased temperature and shifts in patterns of precipitation, is a new challenge to the world of pastoralism. The possible negative effect of global climate change is an added dimension to the already prevailing crisis in the pastoral livelihood system, which is substantially driven by non-climatic factors of internal and external pressures of change such as population growth, bad governance and shrinking rangeland areas lost to competing activities.

Therefore, the impact of climate change and variability on African pastoral systems could be severe because of the prevailing circumstances. The scale of climate change impacts may differ across regions, but it is quite crucial for research to focus on these fragile dryland environments in order to inform sound policy decisions.

There is a dearth of systematic and robust empirical investigation on the economic impact of climate change on pastoral production systems. The burgeoning literature on the assessment of the economic impact of climate change in agriculture has almost exclusively focused on crop agriculture and mixed farming systems. There is a critical gap of more systematic inquiry into the nature and economic impacts of adaptation practices, the challenges of local adaptive responses, and long-term welfare implications of specific adaptation options in the context of traditional pastoralism.

In addition to individual adaptation options ascertained at household level, there are several public measures and scattered efforts that can be identified in response to the need to enhance pastoralist adaptive capacity to climate change. Many are constructive and sustainable. However, many can be destructive to the extent that they may increase pastoralist vulnerability. Thus, a critical and systematic appraisal of adaptive measures in pastoral areas is essential.

The Centre for Climate Change Studies (CCCS) of the University of Dar es Salaam with support from the Open Society Initiative for Eastern Africa (OSIEA), is implementing a project on *Building Knowledge to Support Climate Change Adaptation for Pastoralist Communities in East Africa*. The project is designing and testing approaches and mechanisms that will build and strengthen institutional capacity in five countries of East Africa's Arid and Semi-arid Lands (Kenya, Sudan, South Sudan, Tanzania and Uganda). The research component is trying to answer the key question: how are the pastoral communities responding to climate change impacts and to what extent is the pastoral knowledge system enhancing resilience? The research also intends to contribute to policy debates/dialogues regarding the contribution of pastoralism to the five East African political economies. This book, which is an initial product of this worthwhile Project, is therefore, a very welcome input to the continuing debate and dialogue on the efficacy of traditional pastoralism.

Sarah Mukasa,
Deputy Director,
OSIEA, September 2016

About the Authors

Editors

Pius Zebhe Yanda is a Research Professor at the Institute of Resource Assessment, University of Dar es Salaam. He is also the Director of the Institute of Resource Assessment as well as the Centre for Climate Change Studies of the University of Dar es Salaam. He has been a Lead Author of the IPCC Fourth Assessment Report and Special Report on Extreme Events (SREX Report) of 2013. Pius Yanda is also a member of the Joint Scientific Committee (JSC), a Vice Chair of the IPCC and a Member of the IPCC Bureau. He has published extensively on issues of climate change, environment and natural resource management. E-mail: pyanda@gmail.com.

   Claude Gasper Mung'ong'o is currently an Associate Research Professor at the Institute of Resource Assessment, University of Dar es Salaam. He holds a PhD degree in Human Geography from Stockholm University, Sweden, and has more than 20 years of research experience in Environmental Sociology (including climate change), Pastoral Livelihood Analysis and Food Security, and general rural development. He has been involved in numerous socio-economic studies as a member of various natural resource use studies. He teaches several post-graduate courses at the University of Dar es Salaam, including the *Political ecology of natural resource management, Integrated social survey methods, Impacts of climate change on human and ecosystems,* and *Pastoralism and climate change governance.* He has published widely in his areas of research interest, including editing such books as *Geophysical and Socio-Political Dynamics of Environmental Degradation and Conservation in Kondoa District, Tanzania; People's Perceptions and Community Response to Climate Change and Variability: Selected Cases from Tanzania, etc.* E-mail: claude_gasper@yahoo.com.

Contributors

EGERO, Anthony is a PhD holder, educator, a systems ecologist, a researcher and grants manager. His research interests are in strengthening rangeland health monitoring, restoration and management, and climate change adaptation and mitigation. He has conducted historical research on the climate futures for north-eastern Uganda, vegetation phenology, and information systems in semi-arid areas of Africa.

JAKISA, Emmy Saviour is an environmentalist and a forester. He is generally interested in research that improves community livelihoods through plant conservation and management. His research interests are particularly in the areas of agroforestry, plant ecology, biodiversity conservation, and climate change. He is a member of *NatureUganda*.

KATEKA, Adolphine is an economist with a PhD in Human Geography from Stockholm University, Sweden. She also has a Master's degree in Business Administration (MBA) from the Stockholm School of Business Studies and holds a Bachelor of Arts in Economics from the University of Dar es Salaam. She took extensive studies and research in the area of climate change and sustainable development at Sodertorns University's School of Life Sciences, and at the Stockholm Resilience Centre, Sweden. She is currently a senior lecturer and researcher at the Centre for Climate Change Studies at the University of Dar es Salaam. Adolphine Kateka is a member of the Tanzania Commission for Science and Technology (COSTECH) Advisory Committee on Research and Development. She is a member of the Western Indian Ocean Marine Science Association. She also sits on the Green Climate Finance (GCF) Board as adviser to the board. In addition, she is a member of the Advisory Committee of the African Group of Negotiators (AGN), who are part of the United Nations Framework Convention on Climate Change (UNFCCC), a Subsidiary Body for Scientific and Technological Advice (SBSTA). Adolphine Kateka has published extensively in her areas of expertise. Her publications include book chapters such as "Climate Change Financing: Experience from Tanzania" in *Public spending on climate change in Africa: Experiences from Ethiopia, Ghana, Tanzania and Uganda* (2016); "Poverty-illegal fishing nexus in the Lake Victoria Fisheries" in *Accelerating economic growth and food security in Africa: Contribution of capture and aquaculture fisheries* (2014); and "The economics of value chains: A case of the Lake Victoria Nile Perch" in *The economics of food production in a changing environment*(2012).

KYOBUTUNGI, Angella holds a degree in Environmental Science from Makerere University, Uganda. She is currently working as a Field Technician in the National Agricultural Research Organization at Ngeta Zonal Agricultural Research and Development Institute. Her research interests include biodiversity and ecosystems at risk, climate change adaptation and mitigation, and intervention approaches in agriculture.

LIWENGA, Emma Teresa is a Senior Research Fellow at the Institute of Resource Assessment (IRA) as well as the Centre for Climate Change Studies (CCCS) of the University of Dar es Salaam, Tanzania. She holds a PhD degree from Stockholm University in Sweden and has considerable experience in conducting applied research in analysis of agricultural systems, food security and rural livelihoods as linked to natural resource management. Emma Liwenga has further been engaged in conducting as well as coordinating climate change adaptation research, particularly using Participatory Action Research, to strengthen the adaptive capacity of rural communities to adapt to climate variability and change within the agricultural innovation system. The key articles published include: "Rainfall, Food Security and Human Mobility: Case Study - Tanzania" produced in collaboration with CARE & UNU-EHS. She has also been involved in reviewing climate change adaptation policies on agricultural systems at East Africa regional as well as at national levels. Areas of research interest include food security, gender, climate change adaptation and natural resource management. E-mail: liwenga99@gmail.com.

MABHUYE, Edmund is currently working at the Institute of Resource Assessment, University of Dar es Salaam. He is an expert in ecosystems and natural resource management and climate change mitigation and adaptation. He has conducted various research linking natural resource management and community livelihoods in the context of climate change. He has worked in various higher learning institutions, including the University of Dodoma. E-mail: emabhuye@gmail.com.

NANDOZI, Carolyn, who has a Master's Degree, is a climate data analyst, educator, health advocate, researcher and technologist. Her research interests are in promoting and engaging lifelong mobile learning for marginalized farming and health communities of Uganda towards social justice, boosting livelihood development, and climate change adaptation and mitigation. She has conducted a study on downscaling climate change in Uganda while contextualizing General Circulation Models to particular crops grown by small scale farmers at household level. She is currently assessing likely climate change impacts on different sectors for policy and decision making recommendations regarding climate change adaptation and mitigation plans. She is an active member of *NatureUganda*. Nandozi has published several papers focusing on agriculture, climate change, health and ICT areas.

ONYANGO, Paul is a PhD degree holder, social scientist and lecturer at the Department of Aquatic Sciences and Fisheries and the Centre for Climate Change Studies of the University of Dar es Salaam, Tanzania. Dr Onyango's research focuses on institutional aspects of governance, challenges in poverty alleviation, fisheries management and decision-making processes among communities and adaptation and mitigation to climate change. He has been investigating the dilemmas authorities face in confronting poverty and managing small-scale fisheries. He is also investigating adaptation mechanisms of communities as well as pastoralism as a production system. Some of his latest publications include: "Institutions and Co-Management in East African Inland and Malawi Fisheries: A Critical Perspective" in *World Development Journal* (2015); "Governance of Tenure in the Lake Victoria" in *Land Tenure Journal* (2013); *Poverty in Lake Victoria Fisheries – Understanding Governability* a book published by Springer (2013), "Climbing the Hill: Poverty Alleviation, Gender Relationships, and Women's Social Entrepreneurship in Lake Victoria, Tanzania" in *Maritime Studies Journal* MAST (2011), and "Reconfiguring poverty: the wickedness perspective" in the *Journal of Tropical Hydrobiology and Fisheries* (2010). Dr Onyango is also a co-author of "Freedom and Poverty in the Fishery Commons" in the *International Journal of the Commons* (2010); and "Assessing poverty in small-scale fisheries in Lake Victoria, Tanzania" in the *Fish and Fisheries Journal* (2010), to mention just a few.

TABUTI, John Robert Stephen is a professor with the Department of Environmental Management, Makerere University, Uganda. John Tabuti is a PhD degree holder, educator, ethnobotanist and researcher. His interest in ethnobotany is focused on identifying, promoting and conserving useful plant species for climate change mitigation. He has conducted research covering plant ecology, tree propagation and seedling establishment. He is currently leading a regional capacity building project on climate change mitigation focusing on REDD+. Apart from his academic work at Makerere University, he is also a Council Member and Fellow of the Uganda National Academy of Sciences (UNAS) and a Corresponding Member of the Royal Academy of Overseas Sciences of Belgium. Email: jtabuti@caes.mak.ac.ug.

PART I

Introduction

"Pastoralism is not just a question of one animal...the human being... following another...livestock...; people need to know that the pastoralist is a hero who has overcome adverse conditions of nature to make a viable livelihood...."

Ali Wario, Assistant Minister, Ministry of Special Programmes, Office of the President, Republic of Kenya

CHAPTER 1

Pastoralism for a Sustainable Future in East Africa – An Introduction

C.G. Mung'ong'o and P.Z. Yanda

Institute of Resource Assessment and Centre for Climate
Change Studies
University of Dar es Salaam, Tanzania

An Overview

As has been reported by the IPCC (2001), changes in temperature and rainfall patterns are widely predicted, with many arid and semi-arid parts of the developing world becoming even hotter and dryer. These changes will both directly affect crop yields and will produce changes to ecosystem distributions and species ranges. This will affect the livelihoods of many communities through declining food security and constraints with variability of many livelihood activities, including the pastoralist production systems and communities mostly dependent on livestock keeping.

The degree of vulnerability of African natural and human systems is likely to be significant given the trends of the observed and projected climate change impacts by the IPCC (2001). Dependence on climate sensitive sectors such as nomadic and semi-nomadic animal husbandry and rain-fed agriculture exposes Africa to be more severely affected than other regions in the world.

The East Africa region, which includes the two Sudans, is home for thousands of pastoralists who herd their livestock in the semi-arid to arid areas of the region. Rainfall seasonality affects forage availability, livestock production and ultimately the livelihoods of these people. East African rainfall is bimodal, but is characterized by uncertainty, both spatially and temporally.

Some climate analysts suggest that there will be highly differential impacts of climate change in East Africa to the middle of the twenty-first century. Parts of East Africa will become drier, with considerable reduction in the length of the growing season, while other areas, including southern Kenya and northern Tanzania, may become wetter, with increases in the length of the growing season. Such changes will surely make fundamental changes to ecosystem structures and functions. These in turn will affect both human land-use and livelihoods and have the potential to make both human and animal populations more vulnerable.

Despite these challenges, there are still noticeable gaps related to climate change science issues across scales and levels. Climate change knowledge is not well collated and communicated to communities which depend on climate-reliant production systems such as pastoralism. There is also inadequate coordination and implementation of climate related policies and strategies. The chapters in this book are, therefore, concerned with some of these issues and the consequences that some of the development initiatives have historically had on the pastoral communities in the East African region. The book is divided into five related parts as discussed below.

Understanding the Pastoral Production Systems

Pastoralism is a livestock production system that has been practiced in many regions of the world for centuries (WISP, 2007) and is based on extensive land use and often some form of herd mobility. Currently, extensive pastoralism occurs on about 25% of the earth's land area, mostly in the developing world, from Africa to Asia in the drylands where intensive crop cultivation is physically not possible (FAO, 2001).

Worldwide, pastoralism supports about 200 million households and nearly a billion head of animals including camels, sheep, cattle, and smaller livestock that account for about 10% of the world's meat production (FAO, 2001). Pastoralism is globally important for the human populations it supports, the food and ecological services it provides, the economic contributions it makes to some of the world's poorest regions, and the long-standing civilizations it helps to maintain.

Global changes are putting great pressures on pastoralism worldwide. These pressures are associated with human population growth, economic development, and land-use changes. These developments are challenging policy makers to sustain and protect these invaluable social, cultural, economic, and ecological assets worldwide. The nature and characteristics of pastoralism in the East African region is discussed in **Chapter 3** of this book.

Vulnerability and Responses to Climate Change Shocks and Stresses in Pastoral Communities

The distribution of the pastoral peoples of tropical Africa has been illustrated by Hans Jahnke (1982). The areas used constitute the continent's arid and semi-arid rangelands. In these eco-climatic zones, the rainy season varies from a few weeks to three to four months; growing time of forage plants varies from less than 90 days in northern fringes of the Sahara to 180 days in the savanna zone (Jahnke, 1982). Potential evapo-transpiration is extremely high: 1500 to 2500mm per year, in some places almost always exceeding actual rainfall. Such high evapo-transpiration demands result in moisture deficits which drastically reduce duration of plant growth and therefore green forage availability.

Furthermore, rainfall variability, both in space and time, causes uneven and unpredictable levels of forage productivity. For example, in northern Kenya, during the drought years of 1968 to 1976, mean annual rainfall was approximately 50mm. In contrast, rainfall in 1977 was over 350mm, which was reflected in greatly increased plant production (Funk et al. 2005). Even more important is the spatial distribution of rainfall, especially in the rangelands of East Africa, where rainfall is produced by individual storms. This has important influence on how the range resources are used by nomads. The dimensions of vulnerability to climate change and such variability of the East African pastoral production systems is discussed in **Chapters 4** and **5** while the role of culture and gender in shaping rangeland management and adaptation in East Africa is discussed in **Chapter 6.**

Pastoral Knowledge Systems and Resilience to Climate Change

The functioning of the ecological system is dependent on complex interrelationships between people, animal life, plants and the physical environment. Large expanses of savanna, which are known to be some of the most productive ecological systems in terms of plants and animal biomass, are sustained through a dynamic balance of human activity - grazing and burning - and a broad diversity of grazing wild herbivores and predation through carnivores. Pastoral nomadism, which forms part of the functioning of this ecosystem, is a biological necessity for human survival in this environment.

The ability of nomadic people to survive in these marginal lands is attributed to their opportunistic mobility and diversified livestock husbandry. Drought is not a new phenomenon to the pastoral nomads.

Historically they have suffered numerous such disasters. There is, however, serious concern that these drought strategies are breaking down. Today, more pastoralists are seeking outside help than ever before. This has been graphically demonstrated in Ethiopia, Somalia, Kenya and some Sahelian countries where dispossessed nomads sought survival in refugee camps, relief and urban centres. In this monograph, the pastoral nomadic strategies for mitigating drought effects, including the ecological and social significance of these strategies and the conditions that contribute to the gradual breakdown of nomadism are reviewed in **Chapters** 7 and **8.**

Policy Concerns in East Africa in Relation to Pastoralists

Both during the colonial and postcolonial eras, the attitude of governments towards pastoralism has ranged from outright hostility to benign neglect. When governments have intervened in pastoral areas, the result has been failed projects informed by imperatives that are totally inconsistent with the reality of the dry lands (Baxter, 1991). However, there are also positive developments. Kenya and Ethiopia have taken important steps in recognizing pastoralism in policy statements and documents. Kenya has established the Ministry of Northern Kenya and Other Lands intent on supporting pastoralism, while Tanzania's newly created Ministry of Livestock Development and Fisheries has a Department of Pastoral Systems Development.

Protecting the right to grazing land for pastoralists and the right not to be displaced from their lands are only given in Ethiopia, although certain provisions in other sectoral policies and laws undermine these guarantees. While there are provisions to protect communal lands and customary pastoral land-use practices in Kenya, Tanzania and Uganda, failure to ensure adequate implementation and/or contradictory legislation has led to substantial alienation of pastoral lands (Bonfiglioli, 1992; Mattee and Shem, 2006; Lugoe, 2011). Evictions of pastoralists from their ancestral lands are increasingly an issue as governments pursue policies promoting foreign investment in commercial agriculture or wildlife-based tourism (Brockington, 1999; Sulle and Nelson, 2009).

Policies on economic development in East Africa mostly focus on commercialization of agriculture and livestock, and as a result have proved unhelpful to pastoralists. The tendency in policy stipulations is to seek to modernize pastoralism using the same arguments and interventions that have failed to deliver development to pastoral areas for many years (Baxter, 1991). In addition, the endemic conflict that characterizes many pastoral areas has focused the attention of governments and other actors on these regions. However, policies in this field (especially cross-border policies) do not yet sufficiently address the interests and issues of pastoralists. **Chapter 9** of this book addresses some of these important issues.

Contribution of Pastoralism to National Economies

Many policymakers in East African countries have had a blind spot regarding pastoralism, in particular its contribution to economic growth. As shown in **Chapter 10**, which provides an in-depth analysis of the economics of pastoralism in the East Africa region, pastoralism is estimated to be worth USD $800 million a year in Kenya alone. Moreover, over 90% of meat consumed in East Africa and more than 50% of the milk produced comes from pastoral herds.

Existing data significantly undervalues the total economic value of pastoralism to national and regional economies. Statistics on nationally produced goods and services only reflect pastoralism's direct economic *values* (products like milk, meat, livestock, hides, leather and non-timber forest products, directly sold or sold as inputs to national and international supply chains). Behind this there is also the (indirect) added value of the herd as a form of insurance, savings and risk management, and the development of social capital (absence of conflict), the direct employment of up to 20 million East Africans, skill development and benefits to the tourism industry. And then there are the environmental values, such as nutrient recycling, maintenance of pasture productivity and biodiversity, tree regeneration, maintenance of natural ponds and carbon sequestration for both climate change mitigation and adaptation. Finally, **Chapter 11** provides a case study of social stratification in three pastoral communities in one district in northern Tanzania.

References

Baxter, P.T.W. (1991). "Introduction." In: *When the grass is gone: Development intervention in African arid lands*, edited by P.T.W. Baxter. Uppsala: The Scandinavian Institute of African Studies, Seminar Proceedings No. 25, pp. 7-26.

Bonfiglioli, A.M. (1992). Pastoralists at a crossroads: Survival and development: Issues in African pastoralism. Nomadic Pastoralists in Africa Project (NOPA) UNICEF, UNSO, Nairobi.

Brockington, D. (1999). Conservation, displacement, and livelihoods: The consequences of eviction for pastoralists moved from the Mkomazi Game Reserve, Tanzania. *Nomadic Peoples* (NS) 3 (2): 74-96.

FAO (2001). Pastoralism in the New Millennium. FAO Animal Production and Health Paper Number 150. Also available at: http://www.fao.org/DOCREP/005/y2647e/ y2647e00.htm# toc.

Funk, C., Senay, G., Asfaw, A., Verdin, J., Rowland, J., Michaelsen, J., Korecha, D., and Choularton, R., (2005). Recent drought tendencies in Ethiopia and equatorial-subtropical eastern Africa. Washington, D.C.: U.S. Agency for International Development. ftp://chg.geog.ucsb.edu/pub/pubs/ FEWSNET_2005.pdf.

IPCC (2001). The regional impacts of climate change: An assessment of vulnerability. http://www.ipcc.ch.

Jahnke, H. (1982). Livestock production systems and livestock development in Tropical Africa. Kiel: Kieler Wissenschaftsverlag Vauk.

Lugoe, F. (2011). Aligning and harmonizing the livestock and land policies in Tanzania. ESRF Discussion Paper No. 35. Economic and Social Research Foundation, Dar es Salaam.

Mattee, A.Z. and Shem, M. (2006). Ambivalence and contradiction: A review of the policy environment in Tanzania in relation to pastoralism. IIED Issue Paper No 140. London: International Institute for Environment and Development.

Sulle, E. and Nelson, F. (2009). Biofuels, land access and rural livelihood in Tanzania. London: IIED.

World Initiative for Sustainable Pastoralism (WISP) (2007). "Change of Wind or Wind of Change?" Climate Change, Adaptation and Pastoralism.

CHAPTER 2

A Photographic Introduction to Pastoralism in Tanzania
by Vedasto Msungu, photographs taken in 2016 in
northern Tanzania

Cattle approach the middle of a pool to circumvent mud. This pond is a remnant of the Mvomero road manufacturing.

Cows eat tree leaves fallen on the ground due to lack of pasture.

Cattle seek water in Mvomero.

Cattle seek water in Mvomero.

Extreme drought in Mvomero.

A large trench was dug in the village of Kambala in Mvomero district as a border area between herders and farmers as a means of resolving conflicts between them.

An auction is attended by various businessmen in Makongeni Village in Mvomero District.

An auction is attended by various businessmen in Makongeni Village in Mvomero District.

An auction is attended by various businessmen in Makongeni Village in Mvomero District.

Some people use the opportunity of extreme drought to build a watering place for livestock and charge Tsh. 100 per animal in Mvomero.

Best professional cattle at Ovenco Ranch in Mzeri Village, Handeni District.

Best professional cattle at Ovenco Ranch in Mzeri Village, Handeni District.

An auction at Tanga Region is held at Kibirashi Village.

An auction at Tanga Region is held at Kibirashi Village.

An auction at Tanga Region is held at Kibirashi Village.

Maasai pastoralists deal with the drought caused by climate change by building watering places.

Maasai pastoralists deal with the drought caused by climate change by building watering places.

Goats and calves are fed at home, enabling children to go to school.

Climate change in Kilindi has caused pastoral communities to stop moving with their families and hence started building permanent housing.

Houses for pastoral communities.

Pastoralists now build permanent houses of tin for water harvesting due to drought caused by climate change.

Modern houses are built by pastoral communities for permanent residence.

Pastoralists cope with drought by keeping the grass on the area culled olili. On the left of the image, the grass is not conserved for drought and on the right it was well conserved during the summer, offering children the opportunity to go to school.

The conserved area is used to feed livestock (weak calves and cows) during the summer, while other breeds are moved away in response to climate change.

Pastoral communities with modern houses gain information by using solar panels to energise broadcast equipment like televisions.

Pastoral communities with modern houses gain information by using solar panels to energise broadcast equipment like televisions.

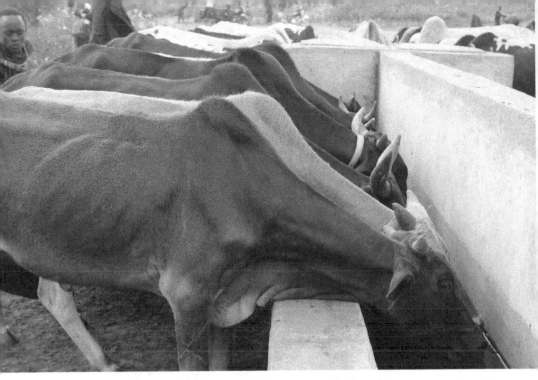

Watering places are built to cope with drought.

During extreme drought, a man digs for water on a dry river in Kiteto District.

Rivers that once flowed throughout Kiteto District have now dried due to drought caused by climate change.

Infrastructure built by stakeholders tackles drought at Endonyongijape Village in Simanjiro District as a way to get water during extreme drought

A Maasai family at Taleta Village in Simanjiro District carries bricks to build their home. Previously such families did not build permanent houses, but due to climate change they have begun building permanent homes.

Maasai women struggle to get water for their families in Simanjiro District, coming home at dawn and returning in the evening or during the night.

Water infrastructure was built to support families, but there is no water due to extreme drought.

Water infrastructure was built to support families, but there is no water due to extreme drought.

Ants are one sign of extreme drought.

Ants are one sign of extreme drought.

Extreme drought has dried up the pool of water in Endonyongijape Village in Simanjiro District.

Some Maasai have started to buy the best bulls to improve their livestock and handle climate change.

Maasai women participate in different associations to increase their income and meet the challenges of climate change. Previously, when nomads, it was not easy to gather them together.

Maasai women participate in different associations to increase their income and meet the challenges of climate change. Previously, when nomads, it was not easy to gather them together.

Symptoms of a severe drought in Laanai Village of Simanjiro District in Manyara Region.

Maasai women use donkeys to seek water over long distances during drought.

Maasai women use donkeys to seek water over long distances during drought.

A group of goats at Laanai Village of Simanjiro District in Manyara Region.

CHAPTER 3

Understanding the Pastoral Production System of East Africa

Abdelaziz Gaiballah and Nancy Ibrahim Abdalla

Sudan University of Science and Technology, Range Science,
Sudan

Introduction

Pastoralism is a land-use system associated with the use of natural rangelands by pastoralists for whom livestock raising is an economically and culturally important enterprise. Characteristically, pastoralism has remained the main livelihood practice in the arid and semi-arid areas of eastern Africa where animals have represented not only economic assets, but also social value. When different authors address issues of pastoralists it is obvious that there is an expression of different views about the context of using related terms and about what the terms exactly mean. In the context of this discussion, the term pastoral production system is being used interchangeably with pastoralism, since pastoralism is the system that encompasses humans, rangelands and herds as three pillars that are interlinking and interacting. These three pillars are mediated by the surrounding environment, including climate change, which affects the system's stability.

According to IFAD (2007), pastoralists are people who live mostly in dry remote areas. They are people whose livelihoods depend on their intimate knowledge of the surrounding ecosystems, and practice nomadic pastoralism as a survival strategy in looking for pasture and water. Historically, the issue of settlement of nomadic pastoralists has raised much debate. Many decision makers think that settlement is the key solution to ensure pastoralists' development.

However, many things remain unclear about these issues. For example, when we talk of settlement of pastoralists, what do we really mean? Is it to be for people themselves or for their livestock or about how to synchronize traditional pastoralism and modern ways of living? What type of settlement status are we talking about: total settlement? Semi-settlement with satellite villages in the varied ecosystems? Or transhumance like that of Spain?

As noted by Bennett (1984), post-independence governments eagerly tried to modify the production regimes of pastoralists toward sedentary livestock production, or simply to reduce the amount of land available as pasture to migratory groups, in an attempt to encourage more intensive use and management of resources. Projects of this nature which were continuations of colonial experiments, rarely fulfilled expectations, and in many instances they ended up disrupting pastoral systems and the relationships between pastoralists and their cultivator neighbours and associates.

It is commonly agreed that pastoralism is facing increasing challenges which make it a difficult way of living and a system at a cross roads, and it also raises many questions about its sustainability. However, mobile livestock is still the existing and time tested practical way of practicing pastoralism in the arid and semi-arid areas of eastern Africa. Pastoralism still delivers a major contribution to the economy, food security and environmental management for East African countries and with no foreseen practical substitute yet. This chapter attempts to provide a critical review of the status of knowledge about the pastoral production systems in East Africa within the context of climate change and the future. The chapter is aimed at sharing knowledge and informing policy dialogue towards sustainable development of pastoralism in the region.

Understanding Rangeland Dynamics and their Variability in Semi-Arid Lands of East Africa

Arid and semi-arid lands are areas falling within the rainfall zones of 0-300 mm and 300-600 mm, respectively (FAO, 1987). They are characterized by annual and erratic rainfall (Sullivan and Rohde, 2002). The spatial and temporal variations of rainfall determine the density and composition of vegetation (Hodgkinson, 1992; Coppock, 1993). Many sources defined rangelands as those marginalized lands not suitable for cultivation, although this could be considered an advantage in the sense that pastoralism is making use of low productive areas, it should also be remembered that rangelands dominate but are not necessarily restricted to dry areas. Thus, for example, the summer grazing domains in most

East African countries are areas where feed and water are available during the dry seasons. According to Zaroug (2000), rangelands form a huge natural resource. They constitute various types of grazing lands, which vary from open grasslands to seasonal water courses, flood plains, river banks and associated islands. They also include woodlands, hills and mountain slopes.

According to Behnke (1992), African rangelands are ecologically heterogeneous with a variety of spatial scales. Local heterogeneity is usually expressed in patchy distribution of pockets of high and low range influenced by topography and soil types. This results in spatial and temporal differences of vegetation composition and stages of growth that enable livestock to feed at the best times and best stages of plant growth, making mobility an advantage. On the other hand, the ecology of arid and semi-arid rangelands is not at equilibrium due to rainfall variability and unpredictability. This makes most of the concepts used to drive classical rangeland management, like carrying capacity and the way rangeland conditions are usually assessed, inappropriate to the management of grazing systems that are not at equilibrium.

Most techniques of rangeland management were developed under the paradigm of increasing and sustaining livestock production by decreasing the inherent variability associated with rangelands and grazing which do not suit the nature of African pastoralism. These traditional rangeland management techniques reduce rangeland heterogeneity by favouring the most productive, most palatable forage species for domestic cattle (Fuhlendorf and Engle, 2001) without considering other environmental factors. This situation greatly affected the understanding of development agents, professionals and decision-makers working with pastoralists on the interventions and support needed for appropriate rangelands management.

Mobility and Herding Practices

Mobility has been the common denominator of pastoralism wherever it has been practiced. Mobility which is the main herding practice in arid and semi-arid areas is essential for response to different levels of variability. It does not vary greatly among the East African countries. Throughout history, herding practices have remained almost the same with mobility being at the heart of their success. Mobility has been used as a way of managing rangeland resources and has constituted the best means of sustainable use of the resources of these areas, while ensuring access to the best feed. This overview points to the fact that herding practices and the decisions pastoralists take to manage natural resources are not chaotic. They have been inherited and repeatedly tested for quite a long time. Some of the major herding practices are discussed below.

Herd splitting is among the popular practices applied, whereby pastoralists divide their livestock into separate herd types, by sex or age, in order to take specific types to areas where pastures suit them best. Another practice is the keeping of large herds as a means of facing hard conditions that may lead to loss of numbers of heads but not of all.

Keeping of indigenous breeds is another common practice adopted by livestock keepers. Indigenous breeds are adaptable and capable of living in the conditions of the mobility range, which constitute the only way to make sustainable use of rangelands under arid and semi-arid areas (Lane, 1988). Different types of livestock use different types of rangelands based on whether stocks are grazers, browsers or both. For example, the East African zebu, which is the type of cattle dominantly raised by pastoralists, is able to lower its metabolic rate under heat stress with the capacity to regain the lost weight due to lowering of metabolic rate (Swallow, 1993).

As already pointed out above, mobility is the other important practice used by East African pastoralists. Table 2.1 shows the mobility patterns practiced in the different countries of East Africa. Both Table 2.1 and Figure 2.1 illustrate the different directions of movement based on the nature of resource distribution in each country. Whatever the distances or directions taken, mobility is in principle based on the fact that herders move to better herding conditions or to avoid unfavourable conditions such as diseases, conflicts, shortage in forage or water, *etc.*

Figure 3.1: Map of main livestock routes in East Africa

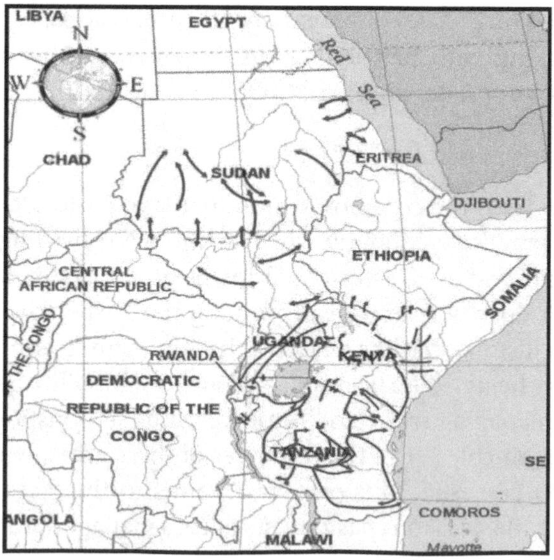

Those who think that mobility is not an appropriate practice consider it a cause of conflict and at the same time a kind of livelihood style that makes impossible any provision of social or development services such as education. They also consider it a livelihood pattern that is no longer coping with the modern way of living and no longer part of the national development priorities. Pastoralists are also accused of keeping herds in excess of the carrying capacity of the environment and hence of promoting environmental degradation. However, experiments done in the Niger (West Africa) in the 1970s (Hesse and Cavanna, 2010) and in the Sudan in the early 1990s (Lamprey and Yussuf, 1981) showed that under dry land conditions, mobile livestock performed better than sedentary ones across a range of indicators, including calving rates, calf mortality and meat production.

In recent years many factors have contributed to reduction of mobility, including urban expansion and irrational land use plans that have mainly increased agricultural lands at the expense of rangelands. In the case of Kenya, expansion of commercial ranches and national parks onto pastoral lands are among these reasons (Schwartz and Schwartz, 1985).

Extensive use of traditional knowledge has enabled pastoralists and their animals to survive in extremely difficult conditions in the arid and semi-arid environments. Pastoralists have developed a very rich experience about their environments and have been able to cope with the harsh conditions. For example, as an insurance mechanism for surviving through periods of stress and maintaining herd structure, they have learnt to keep a sex ratio of 2:3 or 3:4, with females dominating, at the low off-take rate of 6-10% compared to 30% in modern ranches (Niamir, 1990).

They also have developed a very egalitarian social exchange system of loaning, gifting, or sharing resources to assist in compensating loss due to climatic shocks. Rich information has been written about the pastoralist herding practices and their indigenous knowledge. All of which has proved that pastoralists are experienced and skilled resource managers, including that of rangelands and herds. These results confirm the fact that pastoralists know how to use and conserve resources through adoption of environmentally friendly herding with mobility being a necessity and the best way to utilize rangeland resources provided proper land-use plans are in place.

Table 3.1. Profile of the pastoralist production systems, major groups, livestock populations, mobility, rangeland resources and climatic zones.

COUNTRY	PRODUCTION SYSTEMS	ETHNIC GROUPS	LIVESTOCK POPULATION
Sudan	Pastoralism, Agro-pastoralism (De Verdière, 1995; Craze, 2013)	Bega, Shukria, Rufaa, Lahaween, Ahamda, Kababeesh, Misyria (UNDP, 2006)	29 million cattle, 30 million goats, 38 million sheep and 5 million camels.
Kenya	Pastoralism, Agro-pastoralism (Odhiambo, 2006)	Maasai, Turkana, Pokot, Samburu, Garbra, Borana, Rendille, Somali, and Oromo (Odhiambo, M., 2006).	13 million cattle, 25 million goats, 14.9 million sheep, 1.7 million donkeys and 2.9 million camels (KNBS, 2010)
Uganda	Communal pastoral system, Enclosed ranching, Fenced dairy farms, Zero grazing (Mbuza, 1991.)	Bahima, Banyoro, Baganda, Baruli, Itesot, Langi, Karamoja and Karimojong (Kisamba, 1992)	11.4 million cattle, 12.5 million goats, 3.4 million sheep, 3.2 million pigs and 37.4 million poultry (Uganda Bureau of Statistics, 2009)
Tanzania	Pastoralism, Agro-pastoralism and Commercial ranching	Maasai, Barabaig/Datoga (IWGIA, 2012), the Parakuyo and the Sukuma	21.3 million cattle, 15.1 million goats and 5.7 million sheep (URT, 2012)
South Sudan	Pastoralism, Agro-pastoralism (Cately, et al, 2005)	Murle, Jie, Nuer, Dinka, Taposa, and Nyangatom (Abate, 2006)	11.7 million cattle, 12.4 million goats and 12.1 million sheep (Abate, 2006)

MOBILITY TRENDS	RANGELAND RESOURCES	CLIMATIC ZONES
North to south in western, central Sudan, including Butane area, but rather irregular at the Red Sea coast (De Verdière, 1995; Craze, 2013)	Natural grassland and woodland	Arid and semi-arid lands of Sudan comprise appro ximately 70 % of the country's total area (Suliman, and Darag, 1983)
Migration during the two dry seasons between January to March and June and October according to pasture availability (Mutai and Ward, 2000)	Grassland and Erica scrubland, savanna with patches of residual forests	Arid and semi-arid lands make up more than 83% of the country; northern Kenya constitutes most of this area (REGLAP, 2012)
From Mbarara in the southwest to Kaabong in the northeast of the country	Forest, woodland, bush land, grassland and wetlands	Semi-arid areas mostly in Karamoja and parts of Teso, where droughts have been most frequent and most prolonged (Ekiyar, et al., 2010).
Pastoralists to migrate to the central, eastern and southern parts of the country (Sendalo, 2009)	Grassland, dense thickets, woodland gallery forests and seasonal grasslands (Armitage, 1996)	Nearly 80% of the land area in Tanzania is classified as semi-arid (Shem, 2010).
Migration to nutritionally high dry season grazing when flooding, return to higher ground when the rains begin (Abate, 2006)	Natural grasslands, bush and woodland (Abate, 2006)	Tropical climate with wet and dry seasons (http://www.britannica.com/place/South-Sudan)

Although many accuse pastoralism of being a land-use system that contributes to conflicts due to mobility and friction with farmers, in reality this is not a fact. This land-use system has been under practice for a very long time, but incidences of large scale conflicts are very recent occurrences growing with the growing complications associated with transformations in other land uses. Such land uses are linked with the increasing demand of land for urbanization, land grabbing and other irrational land uses. Conflicts on natural resource use are not new or unusual in pastoralist landscapes. They used to happen from time to time, but they also used to be resolved within the traditional administrative systems.

As aptly argued by John W. Bennett, the combination of individual herd ownership and communal grazing land has now been generally understood to be the root cause of the problem (Bennett, 1984). Pastoralism operated without substantial pastureland degradation when pastoralists were free to move over large areas and to work out competitive relationships with one another in their own way, which frequently included raiding and warfare. Cessation of this confrontational regulation through colonial pacification measures removed one important control. At the same time, restriction of movement through closing range areas, drawing political boundaries, or creating block grazing schemes as group ranches reduced the availability of adaptive movement to permit removal of competition, or to seek out suitable pasture in periods of regional drought.

These and other checks and balances in the indigenous system made the institutional combination work. When they were removed, the system collapsed. There was no collective responsibility for the various rangeland resources (Bennett, 1984). It is thus that today's farmer-herder conflicts have become one of the most serious problems facing East African pastoralism, in particular, and the region's peace and security, in general. The situation in the Darfur state of Western Sudan is, perhaps, an extreme case in point.

Pastoralism – Its Importance and Negligence

Many people look at pastoralists as rural communities who live in different countries of East Africa making their livelihood by herding animals and using rangelands, but not looking at pastoralism as a land-use or a production system that supports our economy and contributes to our food security and environmental conservation. This raises a critical question about who owns pastoralism, because this answer indicates who should care about it. The question is also about what is owned, the goods or the services provided or the assets owned by pastoralists themselves? A proper answer will help to change the existing negative attitudes and hence improve the related decisions.

In 1991, for example, Mohamed Salih showed that pastoralists represented 14.1% of the total population of the Sudan and contributed 12% of the GDP and 23.5% of the foreign exchange the country earned from the export of meat and other related products (Salih, 1991). In the case of Tanzania, this system was said to contribute in 2007 up to 13% of Tanzania's agriculture GDP in addition to the fact that it also produced more than 90% of the meat and milk consumed in the country (Mdoe and Mnenva, 2007). In the same year another analyst commented that almost 80% of the livestock in East African countries was raised under this system. It also constituted 20% of Sudan's annual GDP and was a source of all the meat consumed in the country (Fahey, 2007). There are no reliable figures to account for the present situation.

The fact that there are no reliable up to date figures is a clear indication that decision makers pay little attention to pastoralism or to gathering information about pastoralism to correct the accumulated misconceptions and lack of statistics.

Sustainability of the Pastoral Production System

Pastoralism is a system that has humans as one of its main pillars and so it is a dynamic system that is expected to undergo changes. The changes are continually affected by different driving forces that determine the sustainability of pastoralism as a form of livestock raising. The level of challenges is also changing with an exponential increase of the factors that influence the system.

Climatic factors, in this case, are not the only reason that interferes with the herding pattern and praxis. There are also social and economic reasons that may do that. In Kenya, for example, the value of a camel is increasing as the animal now becomes increasingly more important as the droughts in 2005-2006 led to a 70% loss of the pastoral herds (Baird, 2008). The Beja tribes of eastern Sudan started to replace their traditional known type of camel (the *Bushari* that is known to cope with the Red Sea environment) with the *Doauli* type which produces more milk, but is not adaptable to the Red Sea environment. This change arose because there was a growing demand for camel milk by those working in gold mining (Gaiballah and Abdalla, 2015).

On the other hand, climate change and variability are greatly affecting pastoralism as exhibited in the growing scarcity of vegetation, water and animal feed. The power of pastoralism as a self-regulating system depends greatly on traditional tribal management and control, which has been greatly weakened across the different countries (Bennett, 1984). In Sudan, for example, the local administrative system was abolished in 1971, greatly affecting the communal system in western and eastern Sudan where pastoralism is a prominent livelihood (Markakis, 2004).

Two contrasting views about the sustainability of pastoralism have emerged. One view considers pastoralism as a most relevant land-use practice that suits semi-arid areas and contributes significantly to food security and national economics, in addition to assisting in environmental conservation, while the other view considers the existing pastoral production system as no longer able to continue within the prevailing conditions and the expected rhythm of change. The second trend of thinking has been greatly influenced by the tragedy of the commons idea first mooted by Garret Hardin in 1968 (Hardin, 1968). It is also influenced by the widespread view that rangelands are greatly degraded by the overgrazing that is imposed by increasing numbers of livestock. The proponents of this latter view believe that the production system that depends on commercial ranch blocks and grazing rotations can be the solution to secure provision of livestock products.

In reality, the intensive production system requires intensive inputs, and availability of animal feed on the site year-round, while traditional nomadic or transhumant livestock herding makes use mainly of areas where the potential for crop cultivation is limited due to low and highly variable rainfall (IFAD, 2009). It is a low input but high output system that makes a means of living for large numbers of households in arid and semi-arid lands. As this is a different production environment, it differs greatly from other environments and so alternate ways of thinking and solutions are needed. These solutions will not be imported and should be part of the research that suits the real context and understanding of the actual problems.

Sustainability should also consider the integrity of the rangeland, as the resources are for both wildlife and livestock. What is needed is ecological/environmental balance to safeguard and maintain rangeland productivity and dynamism. It is not just about changing and replacing, it is about keeping a balanced relationship between pasture, animals and people, in literature defined as the *triangle of sustainability* (Koocheki and Gliesman, 2005).

In order to optimize the components of pastoralism and to achieve sustainability, there is need to mobilize research and to value the importance of research work in this respect. Research may need to address pastoralism as a livelihood and as a viable production system. One of the areas that need support from decision makers is provision of resources for research in pastoralism as justified by the fact that it is a strategic livelihood issue in East Africa.

Conclusion

Facts and evidence from research show that pastoralism is an environmentally friendly land-use and production system that contributes significantly to the East African countries' economies and ensures food security at the household/boma level. This system has existed for thousands of years and has proved to be adaptive and resilient. As with any other system, it is facing challenges linked to factors such as climate change, irrational land use and other social, political and environmental factors. We all agree that it is an important system that needs to be properly managed and maintained for the benefit of all. What is needed is to ensure sensitization on how to address and sustain pastoralism. There are different views about the efforts needed to attain a clear understanding of how to sustain pastoralism, ensuring that it has social, economic and environmental importance to the East African countries.

Recent research shows that pastoral communities are well adapted to the harsh conditions of the arid and semi-arid areas (Scoones, 1995), making it a resilient system. The growing social, environmental and climatic factors, especially climate change, represent real challenges for pastoralism. There is a clear need to understand pastoralism's components and the mechanism of their interactions and to correct the misconceptions that have led to negative impacts on the system at the level of management or decision making. Misconceptions also arise because the media focuses more on the touristic aspects of rangeland and nomads such as their social celebrations, but rarely reflects on the real problems facing pastoralists and pastoralism. Results oriented action-research and capacity building efforts such as conducting courses, programmes and workshops for different stockholders are expected to improve the understanding of pastoralism and help in achieving the desired objectives.

References

Abate, A.L. (2006). Livestock production challenges in the rangelands ecosystem of South Sudan. Paper presented at the Workshop on Environmental Management Plan for Post-conflict Sudan, Raha Hotel, Juba, South Sudan, and October 31-November 2nd, 2006.

Armitage, D.R. (1996). Environmental management and policy in a dryland eco-zone. The Eyasi-Yaeda basin, Tanzania. *Ambio* 25 (6): 396-402.

Baird, R. (2008). The impact of climate change on minorities and indigenous people. London: Minority Right Group International.

Behnke, R. (1992). New directions in African range management policy. Pastoral Development Network Paper 32c. London: Overseas Development Institute.

Bennett, J.W. (1984). Political ecology and development projects affecting pastoralist peoples in East Africa. Land Tenure Center, University of Wisconsin-Madison.

Britannica. "South Soudan." *http://www.britannica.com/place/South-Sudan.*

Cately, A., Leyland, T. and Bishop, S. (2005). Policies, practice and participation in complex emergencies: the case of livestock interventions in South Sudan. A case study for the Agriculture and Economics Division of the Food and Agriculture Organization. Rome: FAO.

Coppock, D.L. (1993). "Vegetation and pastoral dynamics in the southern Ethiopian rangelands: implications for theory and management." In: *Range ecology at disequilibrium: New models of natural variability and pastoral adaptation in African savannas,* edited by R.H. Behnke, I. Scoones and C. Kerven. Nottingham: Overseas Development Institute; pp.42-61.

Craze, J. (2013). Dividing lines: Grazing and conflict along the Sudan–South Sudan border. Geneva: Small Arms Survey, Graduate Institute of International and Development Studies, pp. 261.

De Verdière, C.P. (1995). Les conséquences de la sédentarisation de l'élevage au Sahel. Etude comparée de trois systèmes agropastoraux dans la région de Filangué, Niger. Thèse présentée pour l'obtention du titre de Docteur de l'Institut National Agronomique Paris-Grignon; 220 pp.

Ekiyar, V., Jumbe, C., Mangisoni, J. and Mkwambisi (2010). Climate change impacts to agriculture in Uganda's semi-arid areas, farmers' means of adaptation: An application of cross-section data. Second RUFORUM Biennial Meeting 20 - 24 September 2010, Entebbe, Uganda.

Fahey, D. (2007). The political economy of livestock and pastoralism in Sudan. IGAD Working Paper No. 06-08. Addis Ababa: IGAD Livestock Policy Initiative.

FAO Committee on Agriculture (Ninth Session) (1987). Improving productivity of dryland areas. Rome: FAO.

Fuhlendorf, S.D. and Engle, D.M. (2001). "Restoring heterogeneity on rangelands: Ecosystem management based on evolutionary grazing patterns." *BioScience* 51 (8): 625.

Gaiballah, A. and Abdalla, N.I. (2015). *Livelihood transformation as influenced by socio-environment changes and its impact on food security among small producers of the Red Sea zone of Eastern Sudan.* Berlin: Trope Tag.

Hardin, G. (1968). "Tragedy of the commons." *Science* 162:1243-48.

Hesse, C. and Cavanna, S. (2010). *Modern and mobile: The future of livestock production in Africa's dry lands.* London: International Institute for Environment & Development (IIED) and SOS Sahel International UK.

Hodgkinson, K.C. (1992). "Elements of grazing strategies for perennial grass management in rangelands." In: *Desertified grasslands: Their biology and management,* edited by G. P. Chapman. London: Academic Press Ltd; pp.77-94.

IFAD (2007). Combating range degradation: The experience of IFAD. IFAD, 31 October.

International Work Group for Indigenous Affairs (IWGIA) (2012). Country technical notes on indigenous peoples' issues. The United Republic of Tanzania. IWGIA.

Kenya National Bureau of Statistics (2010). Kenya population and housing census. Nairobi: KNBS.

Kisamba-Mugerwa, W. (1992). *Rangeland tenure and resource management: An overview of pastoralism in Uganda.* Kampala: Makerere Institute of Social Research.

Kookcheki, A. and Gliessman, S.R. (2005). "Pastoral nomadism, a sustainable system for grazing land management in arid areas." *Journal of Sustainable Agriculture* 25(4): 113-131.

Lane, C.R. (ed) (1988). *Custodians of the commons: pastoral land tenure in East and West Africa.* London: Earthscan Publications.

Lamprey, H. and Yussuf, A. (1981). "Pastoralism and desert encroachment in northern Kenya." *Ambio* X: 131-134.

Markakis, J. (2004). Pastoralism on the margin. London: Minority Rights Group International.

Mbuza, F.M.B. (1991). A systems analysis of milk production systems in Uganda and prospects for technological change. Ph.D. Thesis, University of Melbourne, Australia.

Mdoe, N. and Mnenva, R. (2007). Study on options for pastoralists to secure their livestock. Arusha: TNRF.

Mutai, C.C. and Ward, M.N. (2000). "East African rainfall and the tropical circulation/ convection on intraseasonal to interannual timescales." *Journal of Climate* 13:3915-3939.

Niamir, M. (1990). "Herders' decision-making in natural resource management in arid and semi-arid Africa." *Community Forestry* Note No. 4. FAO, Rome.

Odhiambo, M. (2006). Review of literature on pastoral economics and marketing: Kenya, Tanzania, Uganda and Sudan. RECONCILE, Kenya.

REGLAP (2012). Key statistics on the dry lands of Kenya, Uganda, and Ethiopia. Nairobi: Regional Learning and Advocacy Programme for Vulnerable Dryland.

Salih, M.A.M. (1991). "Livestock Development or Pastoral Development?" In: *When the Grass is Gone: Development Intervention in African Arid Lands,* edited by P.T.W. Baxter. Seminar Proceedings No. 25. Uppsala: The Scandinavian Institute of African Studies.

Schwartz, S. and Schwartz, H.J. (1985). "Pastoralism in Kenya still a viable production system." *Quarterly Journal of International Agriculture;* 24:5-21.

Scoones, I. (1995). "New directions in pastoral development in Africa," In: *Living with uncertainty: new directions in pastoral development in Africa,* edited by I. Scoones. Exeter: Intermediate Technology Publications; pp.1-36.

Sendalo, D.S.C.A. (2009). Review of land tenure policy implications on pastoralism in Tanzania. Dar es Salaam: Department of Livestock Research, Training and Extension, Ministry of Livestock Development and Fisheries.

Shem, M.N. (2010). Impacts of government policies on pastoralist livelihoods in the semi-arid areas of Tanzania. Fortaleza-Ceará, Brazil.

Suliman, M.M. and Darag, A. (1983). "Desertification with special emphasis on carrying capacity and pastoral resources." *In: Pre-assessment of natural resources in Sudan, IES.*

Sullivan, S. and Rohde, R. (2002). "On non-equilibrium in arid and semi-arid grazing systems." *Journal of Biogeography* 29, 1595–1618.

Swallow, B.M. (1993). The role of mobility within the risk management strategies of pastoralists and agro-pastoralists. ILCA.

Uganda Bureau of Statistics (2009). Statistical Abstract. Kampala: Uganda Bureau of Statistics.

UNDP (2006). Nomads' settlement in Sudan: Experiences, lessons and future action. Khartoum: UNDP Sudan.

United Republic of Tanzania (2012). National sample census of agriculture, smallholder agriculture: Livestock sector – National report. Dar es Salaam: National Bureau of Statistics.

Zaroug, M.G. (2000). Country pasture/forage resource profiles, Sudan. Rome: FAO Crop and Grassland Service (CGPC).

CHAPTER 4

The Impacts of Climate Change on the Pastoral System in East Africa

Nancy Ibrahim Abdalla and Abdelaziz Gaiballah

Sudan University of Science and Technology, Range Science,
Sudan

Background

Pastoralism is a land-use system associated with the use of natural rangelands by pastoralists. It is a system that encompasses humans, rangelands and herds as three interlinked pillars. It is also the main livelihood practice in semi-arid areas of East Africa and because it is a natural resource-based land use system, it is also expected to be the most affected by climate change compared to other land uses in the region. Climate change has a large impact on the renewable natural resources of arid and semi-arid areas which are already characterized by high levels of rainfall variability (Awuor, 2011).

Climate change is expected to cause changes to environmental conditions, influencing the characteristics and growth of forests in rural and urban settings, as well as agricultural production and such other ecological impacts of climate change as shifts of vegetation types and associated impacts on biodiversity expansion of arid lands; decline in water quantity and quality; and stresses from pests, diseases, and wildfire (Alavalapati et al, 2011).

Africa is already a continent under pressure from climate stresses and as such it is also highly vulnerable to the impacts of climate change which is a major issue for all the world's rangelands (Howden et al., 2008;

Wei et al., 2008). Because rangelands as part of the ecosystem are very sensitive to variability in the climate and specifically in atmospheric CO^2 concentration, such changes have the potential to influence rangeland quality and composition (Sangeda and Malole, 2014). This effect is mainly through change in the optimal growth ranges for different species, plant composition and species competition (FAO, 2009).

For example, legume species in grasslands and the proportion of browse in rangelands are likely to increase with rising CO^2, in turn affecting carbon sequestration rates in soils and plants (Thornton, et al., 2008). Otherwise, increased temperature increases drought stress and tissue lignification in plants and, consequently, it also affects their digestibility and decomposition rate. On the other hand, increased temperature and lower rainfall lead to a shift in species composition, increase in vegetation flammability, and hence also to increased fire frequency (Sangeda and Malole, 2014).

Climate Change Impacts on Rangelands

In 2008 the Commission on Climate Change and Development (CCCD) reported that the predicted changes in rainfall patterns will result in increasingly scarce, scattered and unpredictable pastures and will also have an important influence on rangeland species composition in both the short- and long-term, primarily through its differential effect on the growth and reproduction of key forage species (CCCD, 2008). Furthermore, changes in the temporal distribution of rainfall could reduce the effectiveness of rainfall through projected increased variation within seasons (fewer, more intense rainfall events) and from year to year (more frequent droughts) (Stokes, et al 2008). Therefore, according to Chaplin-Kramer and George (2013), "climate change has the potential to impact the quantity and reliability of forage production, forage quality, thermal stress on livestock, water demands for both animal needs and growing forage, and large-scale rangeland vegetation patterns".

The combined effects of seasonal temperature and precipitation patterns will influence not only productivity, but also growing season length and plant phonology in rangelands will be strongly affected by climatic variability (Elhag and Walker, 2008). In arid and semi-arid ecosystems with a single rainy season, there is usually a short growth period followed by a long dry season. In these circumstances, pasture areas become greatly reduced, overgrazed (as there is less pasture available) and degraded. In addition, the amount of water available is very limited (Awuor, 2011). Rainfall is more patchily distributed in arid than in mesic systems, and adequate rainfall from year to year cannot be assumed for a given location (Ellis, 1994). It has also been noted, however, that East African rangelands are generally heterogeneous, due to spatial gradients of climate, soils, landscape and disturbance (Coughenour and Ellis, 1993). Temporal and spatial variability in rainfall, in particular, increases heterogeneity of the system.

Scarcer resources, coupled with current levels of demographic growth, are likely to lead to stronger competition between pastoral communities and between these and other groups, possibly resulting in conflict and even violent clashes (CCCD, 2008). Stable degradation of vegetation cover is one of the basic characteristics of the desertification process, and also rainfall variability coupled with overgrazing could cause severe land degradation that has the potential to lead to desertification. More important, it has also been pointed out that degradation of above-ground vegetation can cause an estimated loss of 6 tons of carbon per ha while soil degradation processes lead to a loss of 13 tons of carbon per ha (Woomer, et al., 2004).

IFAD (2007) stated that causes of rangeland degradation include climatic conditions causing drought and human factors leading to the overuse of natural resources. An extended drought can result in the mortality of perennial plants and the switch to an annual-dominated flora (Hein, 2006). Impacts of climate change on water resources could result in increasing incidences of droughts, changing precipitation intensity and runoff, and lower availability of water for irrigation (Alavalapati et al, 2011). They could also result in increased evapotranspiration by crops and natural vegetation, hence leading to more rapid depletion of soil moisture (FAO, 2013).

In addition, climate change is projected to reduce renewable surface water and groundwater significantly in most dry subtropical regions (Jiménez Cisneros et al., 2014). However, the proportion of browse in rangelands may increase in the future as a result of increased growth and competition of browse species due to increased carbon dioxide levels. Table 3.1 below provides a summary of possible climate change impacts on rangeland systems in East Africa.

Table 4.1:   Possible climate change impacts on East African rangeland ecosystems

NATURE OF IMPACTS	ACTUAL IMPACTS
Direct livestock impacts	• Shifts in rangeland vegetation structure or boundaries
	• Changes in forage quality and quantity
	• Changes in length of growing season
	• Changes in livestock productivity and health
	• Changes in water quality and quantity
Indirect ecological impacts	• Changes in soil quality and productivity
	• Disturbances in ecosystem function (e.g. alterations in biogeochemical cycling, incidence of wild fire, etc.)
	• Changes in biodiversity
	• Changes in habitat suitability for wildlife
Indirect socio-economic impacts	• Changes in food production and security (locally, regionally and nationally)
	• Changes in incomes derived from livestock production, and other rangeland inputs
	• Changes in security and resource competition
	• Changes in land use
	• Changes in recreational use of rangelands and alterations in scenic quality

Source: Adopted from WISP (2010).

Climate Change Impacts on Livestock Production

Climate change has essential impacts on livestock. It affects livestock production in multiple ways, both directly through impacts on livestock performance, and indirectly through impacts on the environment, society and the economy (FAO, 2016). The most important impacts are experienced in animal productivity, yields of forages and feed crops. They are also manifested in animal health and biodiversity, as summarized in Table 3.2. In addition, livestock become less resistant to disease (they often die because of the extreme conditions), there is reduced nutrient availability for animals and ultimately livestock production is reduced, which may have impacts on food security and incomes through reductions in the production of milk and meat for smallholders (Awuor, 2011). Marketing of livestock often declines and their price increases immediately after drought because farmers are rebuilding their herds and more likely to withhold animals from sale (Little, 2009). High temperatures reduce feed intake and thus also milk production, and in the tropics lead to energy deficits and decreased cow fertility, fitness and longevity (Parsons et al., 2001).

Different livestock groups have varying degrees of susceptibility to stress conditions such as more frequent and longer periods of drought under a changing climate. For example, cattle known for slower biological turnover are considered more vulnerable to feed shortages during drought than small ruminants and camels (Lesnoff et al., 2012). Droughts may cause overgrazing, mass migration and concentration around pastures and water resources, leading to increased infections with diseases such as Foot and Mouth Disease and pests (WISP, 2010). Both the distribution and the abundance of livestock disease vectors may be affected by changes in rainfall and temperature regimes, as well as by changes in the frequency of extreme events. The feeding frequency of arthropod vectors may also increase with temperatures rising (WISP, 2010).

In various countries in sub-Saharan Africa, 20 to 60% losses in animal numbers were recorded during serious drought events in the past two or three decades. In north west Kenya, for instance, several years of low rainfall have recently resulted in the death of many livestock, and in a major food crisis among the Turkana pastoralists (Commission on Climate Change and Development, 2008). In South Africa, Niang et al. (2014) reported that dairy yields may decrease by 10 to 25% under certain climate change scenarios. The number, distribution and productivity of permanent pastures and water points, which are so critical for livestock survival during the dry season, are bound to decline (Commission on Climate Change and Development, 2008).

Changes in temperature and precipitation affect the composition and diversity of native animals and plants through altering their breeding patterns, water and food supply, and habitat availability (Feng and Hu, 2007). Livestock water consumption will increase with both temperature and drought. The warmer temperatures are expected to increase livestock production costs owing to lower feed intake and increased requirements for energy to maintain healthy livestock at higher temperatures (Alavalapati et al, 2011). But there is still a lack of assessments of livestock production under climate constraints to support policies that aim at improving resilience in the sector (IPCC, 2014). There are a number of factors that increase livestock vulnerability to climate change, especially in semi-arid and arid regions, including rangeland degradation, fragmentation of grazing areas, changes in land tenure, conflicts and insecure access to land and finally markets (FAO, 2016). Table 3.2 below summarizes the impacts of climate change on livestock production as compiled from Thornton et al. (2009) and Niang et al. (2014).

Table 4.2:   A summary of the impacts of climate change on livestock production

IMPACTS	ANIMALS
Variability in rainfall	- Shortages in drinking and servicing water - Diseases • Increased pathogens, parasites and vectors • Changed distribution and transmission • New diseases
Temperature	- Heat stress • Decreased feed intake and livestock yields • Decreased conception rates • Altered metabolism and increased mortality - Diseases • Distribution and transmission through pathogens, parasites and vectors • Decreased resistance of livestock • New diseases - Domestic biodiversity losses
CO2 in the atmosphere	

Source: Compiled from Thornton et al. (2009) and Niang et al. (2014)

FORAGES AND FEED CROPS	LABOUR FORCE AND CAPITAL
- Decreased yields - Decreased forage quality - Changes in pasture composition (species, communities) • Changes in production systems (e.g. from mixed crop-livestock to rangelands)	- Altered human health and resource allocation to livestock - Decreased productivity - Migrations - Conflicts
- Decreased yields - Decreased forage quality - Changes in pasture composition	
- Partial stomata closure and reduced transpiration - Change in pasture composition	

Climate Change Impacts on Communities

Experience shows that those who are living in the arid and semi-arid lands of East Africa have been subjected to repeated droughts linked with famines, as with the ones which occurred in Sudan and in northern parts of Kenya in 1984. This experience indicated that though pastoralists have good adaptive strategies to deal with variability, climate change impacts may still cause significant negative effects because climate has always a direct influence on the lives of human beings. Many people in the world directly depend for their livelihood on natural resources, which in many cases are more vulnerable. This dependence becomes more of a problem in countries that have weak institutional capacity for managing resources (Heltberg et al. 2010). The direct effects of climate change on indigenous people include environmental changes in temperature and precipitation that lead to increasing aridity and drought, floods and increased melting of ice capped mountains and sea ice, rises in sea levels, changes in wildlife habitats, increase in pest and vector diseases, and increase in soil erosion and landslides (Saitabau, 2014).

The world's poor are disproportionately vulnerable to such environmental changes through loss of livelihood assets, dislocation, hunger and famine in the face of climate variability and change (Anderson et al., 2010). Climate change will most likely affect populations through impacts on the necessities and comforts of life such as water, energy, housing, transportation, food, natural ecosystems, and health systems (Alavalapati et al, 2011).

According to Adger et al. (2000), over 250 million people are directly affected by desertification and some one billion people in over 100 countries are at risk. Besides a temperature increase of some 1 to 2.5°C by 2030, it is predicted that during this period, millions of people – mainly those in developing countries – will face changes in rainfall patterns and extreme events, leading to severe water shortages, droughts or flooding. These events will increase the risk of land degradation and biodiversity loss (FAO, 2009). According to Galvin et al. (2004), drought has had very negative effects on people.

Furthermore, grazing lands are being lost due to drought, increasing population pressure and restricted access to land. These problems lead to greater food insecurity and malnutrition among pastoralists. There are also social problems: conflicts increase due to competition over scarce resources during drought years (Awuor, 2011), forcing pastoralists to settle and grow crops and resulting in considerable reduction in grazing lands (Ayantunde, 2011). Pastoralists may soon lose all of their livestock as the effects of drought continue, which in turn will also affect livestock markets (Wandera, 2011). For example, droughts and desertification intensified competition for resources in the Darfur region of western Sudan, exacerbating the volatile political setting and setting off clashes among farming and herding groups. Many households have fled to camps for internally displaced populations (Satti and Castro, 2013).

Climate change will affect the length of growing seasons, and crop and livestock yields, and bring about increased risk of food shortages, insecurity, and pest and disease incidence, exposing populations to greater health and livelihood risks (FAO, 2009). Climate change related disasters not only affect ecosystems, but cause people to relocate either by choice or by force. Some will be displaced within the boundaries of their affected countries (Internal Displacement or ID) and others will cross state borders (Jacques, 2010). Some will be displaced because of sudden-onset hydro-meteorological disasters, such as flooding, hurricanes, landslides, etc. Others will be affected by slow-onset disasters like desertification, rising sea levels and droughts (Jacques, 2010).

Climate change and climate variability are creating increasingly vulnerable conditions in fragile pastoral ecosystems (Oseni and Bebe, 2011). Pastoralists are suffering from the effects of climate change because pastoralists will be among the first groups to lose their livelihoods, as rangelands and water points dry up (Oseni and Bebe, 2011). Climate change represents a new threat and challenge to Tanzania because many households, social groups and regions have a limited capacity to adapt to climate variability and change (Kisanga, 2005).

Sustainability of Pastoralism under Climate Change

Generally, there are two views about the sustainability of pastoralism. One group looks at pastoralism as a system that does not fit with the prevailing lifestyle and cannot continue due to ongoing drastic changes affecting natural resources in addition to the complicated problems linked with mobility. The second group looks at pastoralism as a coping system with social, economic and environmental importance, for which we have no substitute and which is an environment friendly system.

Pastoralist communities are predictably among the groups most affected by climate conditions, and hence will need access to resources and services that help them to cope with imminent disastrous shocks, protect their livelihood assets and increase their resilience (Birch and Grahn, 2015). The impact of predicted weather extremes even raises the question of the sustainability of pastoralism as a feasible livelihood system.

Traditional pastoral systems of resource management have always included application of strong adaptive strategies. Livestock mobility has been one of the major ways in which African pastoralists have historically managed uncertainty and risk from exposure to a restricted resource base and to losses from droughts (Scoones, 1995). Pastoralists have not only herded livestock. They have also practiced other complementary activities, including subsistence farming. It is thus that there are now mixed farmers, sometimes referred to as agro-pastoralists, combining livestock keeping with crop production.

Shrinkage in rangelands and the change in land use have affected social relations between farming and herding societies that were previously mutually reliant (WISP, 2010). These processes have also led to the adoption of opportunistic grazing practices whereby herd sizes and stocking rates are adjusted as the rainy season unfolds to best match available grazing resources (Hazell, 1999), indicating that the production system is a flexible one.

The way that pastoralists use resources takes into consideration management aspects that conserve the resources. Most nomadic groups in East Africa use range management techniques such as pasture rotation and establishment of grazing reserves. These techniques are frequently used to save forage for critical periods. In Sudan, for example, nomads restore some areas to be used during the dry period. The Maasai of East Africa widen their grazing radius and delay entering dry season grazing areas (Jacobs, 1980).

Herd splitting which is a well-known technique is also widely practiced to maintain the long term productivity of the range and to ensure the conservation of rangeland ecosystems (Oxfam, 2008). The raising of different types of livestock gives an opportunity for using different niches. In East Africa, pastoralists raise a mixture of cattle, camels, goats, and sheep. Maintaining a diverse herd has a number of advantages and it represents a critical adaptation measure (WISP, 2010).

Controlled burning is another useful rangeland management practice that is also used by pastoralists to reduce parasites, destroy unpalatable grass species and shrubs, and encourage the growth of favoured species (FAO, 2016). Improved fodder species improves livestock production and therefore increases resilience of both livestock and pastoralists during stress periods such as drought (Oxfam, 2008). Management of disease is done through adoption of preventative measures such as avoidance of areas known to be particularly susceptible to disease, migration and use of hygienic practices such as dipping and/or hand pump spraying.

Other adaptation options include grassland restoration or diversification in composition; agroforestry with fodder trees and legume shrubs that provide alternative feed resources, shade and retain water (FAO, 2016). Some pastoralists also apply adaptation strategies such as the use of emergency fodder in the form of grazing enclosures, culling of weak animals, etc. Pastoral women play a significant role in natural resource management. During drought they harvest supplementary wild foods and other products of market value such as honey (WISP, 2010).

Intervention in terms of active selection of plant species and controlled animal stocking rates is perhaps the most promising management activity to reduce the negative impacts of future climate scenarios on rangelands. At the international level, the pastoral system is also greatly benefiting from the huge global concern and support connected with climate change. This support focuses on maintaining the sustainability of natural resource systems and rangelands. The BRACED funding from DFIFD-UK is, for example, one of the biggest of such international efforts. The project "Building knowledge to support climate change adaptation for pastoralist communities in East Africa" funded by OSIEA is another example of these efforts. This is an opportunity for those working in pastoralism to mobilize research and knowledge, as it is also the best time to work towards influencing policy related issues and decision making processes. Table 3.3 below shows some of the adaptation measures that have been taken in dealing with climate change impacts on the pastoral production system.

Table 4.3 Adaptation measures taken by pastoralists to deal with climate change impacts

CLIMATE INDUCED CHANGE
Altered water availability
Drought
Erosion and soil degradation
Pests, disease, and invasive species
Changes in vegetation composition
Heat stress

Source: Adapted from Thornton et al. (2009) and Niang et al. (2014).

ADAPTATION MEASURES

- Adoption of integrated water resource management
- Introduction of water harvesting and conservation techniques, including hafirs, small size dams, water harvesting systems, boreholes, sub-surface and sand dams.
- Introduction of drought tolerant plant species
- Planting of live shelterbelts and trees to increase shade

- Sustainable management for natural stands in semi-arid areas
- Introduction of heat/drought tolerant, early maturing and high yielding varieties of crops
- Introduction of improved fodder species to increase and improve livestock production
- Introduction of water harvesting and conservation measures
- Enrichment of tree and shrub density through afforestation and reforestation
- Implementation of fire prevention measures
- Availing feed supplementation to livestock

- Implementation of measures to control soil erosion, floods and runoff
- Restoration of sand dunes
- Creation of shelterbelts and live fences
- Implementation of fire prevention measures
- Increased use of trees for shading and reducing wind erosion

- Improvement of veterinary care
- Improvement of pest predictive tools and indicators
- Increased (but cautious) use of biological and other controls
- Adoption of Integrated Pest Management (IPM)
- Improvement of general health and conditions of rangelands

- Sustainable management of rangeland and woodland
- Adoption of ecosystem resource management
- Eradication of invasive species
- Introduction of drought and salt tolerant varieties

- Improvement of microclimate through afforestation and rehabilitation of degraded areas
- Improvement of water supply through water harvesting and conservation

References

Adger, W.N., Benjaminsen, T.A., Brown and K., Svarstad, H. (eds.) (2000). Advancing a Political Ecology of Global Environmental Discourse. London: Centre of Social and Economic Research on the Global Environment, University of East Anglia.

Alavalapati, J., Alig, R.J., Lal, P., Mercer, D.E., Anita, T.M., Stone, E.A., and White, E.M. (2011). Effects of Climate Change on Natural Resources and Communities: A Compendium of Briefing Papers. (USDA). General Technical Report, PNW-GTR-837.

Anderson, S., Morton, J., and Toulmin, C.. (2010). "Climate change for agrarian societies in drylands: Implications and future pathways." In: Social dimensions of climate change: Equity and vulnerability in a warming world, edited by Mearns and Norton. Washington, DC: World Bank.

Awuor, C.B. (2011). Vulnerability in Ethiopia's pastoral communities. Joto Afrika.

Ayantunde, A.A. (2011). Coping with drought in Niger. Joto Afrika.

Birch, I. and Grahn, R. (2015). Pastoralism - Managing multiple stressors and the threat of climate variability and change. Available online: ttp://hdr.undp.org/sites/default/files/birch_isobel_and_grahn_richard.pdf.

Chaplin-Kramer, R. and George, M.R. (2013). Effects of climate change on range forage production in the San Francisco Bay area. DOI: 10.1371/journal.pone.0057723.

Commission on Climate Change and Development (CCCD) (2008). Climate change and drylands. London: International Institute for Environmental Development.

Coughenour, M.B. and Ellis, J.E. (1993). "Climate and landscape control of woody vegetation in a dry tropical ecosystem, Turkana District, Kenya." Journal of Biogeography 20: 283-398.

Elhag, M. and Walker, S. (2008). Impact of climate change and the human activities on land degradation in arid and semi-arid regions. The 3rd International Conference on Water Resources and Arid Environments and the 1st Arab Water Forum.

Ellis, J. (1994). Climate variability and complex ecosystem dynamics: Implications for pastoral development. In: Living with uncertainty: New directions in pastoral development in Africa edited by I. Scoones. London, UK: Intermediate Technology Publication. pp. 37-57.

FAO (2016). Climate change and food security: risks and responses. Rome: FAO.

FAO (2013). Report of the First Meeting of the Plenary Assembly of the Global Soil Partnership (Rome, 11-12 June 2013). Hundred and Forty-Eighth Session, Rome, 2-6 December 2013, CL 148/13, Rome.

FAO (2009). Review of evidence on drylands pastoral systems and climate change. Implications and opportunities for mitigation and adaptation, edited by C. Neely, S. Bunning and A. Wilkes. Rome: FAO.

Feng, S. and Hu, Q. (2007). "Changes in winter snowfall/precipitation ratio in the contiguous United States." Journal of Geophysical Research 112: D15109. DOI: 10.1029/2007JD008397.

Galvin, K.A., Thornton, P.K., Boone, R.B. and Sunderland, J., (2004). "Climate variability and impacts on East African livestock herders: the Maasai of Ngorongoro Conservation Area, Tanzania." African Journal of Range & Forage Science, 21(3): 183–189.

Hazell, P., (1999). "Public policy and drought management in agro-pastoral systems." In: Property rights, risk and livestock development in Africa, edited by N. McCarthy, B. Swallow, M. Kirk, and P. Hazel. Washington, DC: International Food Policy Research Institute, and Nairobi: International Livestock Research Institute, pp. 86-101.

Hein, A.L. (2006). "Climate change in Africa." Cicerone 6.

Heltberg, R., Seigel, P.B. and Jorgensen, S.L. (2010). "Social policies for adaptation to climate change." In: Social dimensions of climate change: Equity and vulnerability in a warming world, edited by Mearns and Norton. Washington, DC: World Bank.

Howden, S.M., Crimp, S.J., and Stokes, C.J. (2008). "Climate change and its effect on Australian livestock systems." Australian Journal of Experimental Agriculture 48, 780–788. doi: 10.1071/EA08033.

IFAD (2007). Combating range degradation: The experience of IFAD. IFAD, 31 October.

IPCC (2014). "Climate change 2014: impacts, adaptation, and vulnerability." Part A: Global and sectoral aspects. Contribution of Working Group II to the Fifth Assessment Report of the Intergovernmental Panel on Climate Change. Edited by Field, C.B., Barros, V.R., Dokken, D.J., Mach, K.J., Mastrandrea, M.D., Bilir, T.E., Chatterjee, M., Ebi, K.L., Estrada, Y.O., Genova, R.C., Girma, B., Kissel, E.S., Levy, A.N., MacCracken, S., Mastrandrea, P.R., and White, L.L. Cambridge, UK, and New York, USA: Cambridge University Press.

Jacobs, A.H. (1980). "Pastoral Maasai and tropical rural development." In: Agricultural Development in Africa, edited by R.H. Bates and M.F. Lofchie. Issues of Public Policy. New York: Praeger; pp. 275-300.

Jacques, B.N. (2010). "Climate change: What about the displaced?" Policy Analyst.

Jiménez Cisneros, B.E., Oki, T., Arnell, N.W., Benito, G., Cogley, J.G., Döll, P., Jiang, T. and Mwakalila, S.S. (2014). "Freshwater resources." In: Climate Change 2014: Impacts, adaptation, and vulnerability. Part A: Global and Sectoral Aspects. Contribution of Working Group II to the Fifth Assessment Report of the Intergovernmental Panel on Climate Change. Edited by C.B. Field, V.R. Barros, D.J. Dokken, K.J. Mach, M.D. Mastrandrea, T.E. Bilir, M. Chatterjee, K.L. Ebi, Y.O. Estrada, R.C. Genova, B. Girma, E.S. Kissel, A.N. Levy, S. MacCracken, P.R. Mastrandrea and L.L. White. Cambridge, United Kingdom and New York, USA: Cambridge University Press, pp. 229–269.

Kisanga, D.R. (2005). Tradition and Science: Environmental Change in East African Mountains. Middletown, Ohio: Geography Department at Miami University.

Lesnoff, M., Corniaux, C., Hiernaux, P. (2012). "Sensitivity analysis of the recovery dynamics of a cattle population following drought in the Sahel region." Ecological Modeling 232:28–39. doi: 10.1016/j.ecolmodel.2012.02.018.

Little, P. (2009). Hidden value on the hoof: Cross-border livestock trade in Eastern Africa. Policy Brief 2, COMESA.

Niang, I., Ruppel, O.C., Abdrabo, M.A., Essel, A., Lennard, C., Padgham, J. and Urquhart, P. (2014). "Africa." In: Climate change 2014: Impacts, adaptation, and vulnerability, edited by V.R. Barros, C.B. Field, D.J. Dokken, M.D. Mastrandrea, K.J. Mach, T.E. Bilir, M. Chatterjee, K.L.

Ebi, Y.O. Estrada, R.C. Genova, B. Girma, E.S. Kissel, A.N. Levy, S. MacCracken, P.R. Mastrandrea & L.L. White. Cambridge, United Kingdom and New York, USA: Cambridge University Press.

Oseni, S and Bebe, B. (2011). "Adapting pastoralism to a changing climate." Joto Afrika.

Oxfam (2008). Survival of the fittest: Pastoralism and climate change in East Africa. Oxfam Briefing Paper, August.

Parsons, D.J., Armstron, A.C., Turnpenny, J.R., Matthews, A.M., Cooper, K., Clark, J.A. (2001). "Integrated models of livestock systems for climate change studies." Global Change Biology, 7:93–112.

Saitabau, H.O. (2014). Impacts of climate change on the livelihoods of Loita Maasai pastoral community and related indigenous knowledge on adaptation and mitigation. Unpublished MS in National Museum of Kenya.

Sangeda, A.Z. and Malole, J.L. (2014). "Tanzanian rangelands in a changing climate: Impacts, adaptations and mitigation." Net Journal of Agricultural Science 2(1):1-10.

Satti, Y.H. and Castro, A.P. (2013). Climate change and the future of onion and potato production in West Darfur, Sudan: a case study of Zalingei locality. Practical Action Publishing.

Scoones, I. (1995). "New directions in pastoral development in Africa," In: Living with uncertainty: new directions in pastoral development in Africa, edited by I. Scoones. Exeter: Intermediate Technology Publications; pp.1-36.

Stokes, C.J., Ash, A. and Howden, S.M. (2008). Climate change impacts on Australian rangelands. Canberra: Society of Range Management.

Thornton, P.K., Van de Steeg, J., Notenbaert, A. and Herrero, M. (2009). "The impacts of climate change on livestock and livestock systems in developing countries: a review of what we know and what we need to know." Agricultural Systems, 101(3): 113–127.

Thornton, P.K. and Herrero, M. (2008). "Climate change, vulnerability and livestock keepers: Challenges for poverty alleviation." Proceedings of the Livestock and Global Climate Change Conference, May 2008, Tunisia.

Wandera, B. (2011). "Livestock insurance in northern Kenya." Joto Afrika.

Wei, Y., Ruiqing, G., Xuebiao, P., and Buju, L. (2008). "The effect of climate change on livestock carrying capacity of Inner Mongolia." In: Multifunctional Grasslands in a Changing World. Vol. 1. Proceedings of the International Grasslands and Rangelands Congress. Guangzhou, China: Guangdong Peoples Publishing House.

Woomer, P.L., Toure, A. and Sall, M. (2004). "Carbon stocks in Senegal's Sahel transition zone." Journal of Arid Environments, 59: 499-510.

World Initiative for Sustainable Pastoralism (WISP) (2010). Building climate change resilience for African livestock in sub-Saharan Africa - a program of IUCN - The International Union for Conservation of Nature. Nairobi: Eastern and Southern Africa Regional Office, March.

PART II

Vulnerability and Responses to Climate Change Shocks and Stresses in Pastoral Communities

CHAPTER 5

Dimensions of Vulnerability to Climate Change and Variability in Pastoral Production Systems of East Africa

Anthony Egeru, Carloyne Nandozi, Angella Kyobutungi and J.R.S. Tabuti

Department of Environmental Management, Makerere University, Uganda

Introduction

Vulnerability has multiple meanings (Luers, 2005; Adger, 2006; O'Brien *et al.*, 2007; Vogel *et al.*, 2007; Ribot, 2011; Bassett and Fogelman, 2013), but in the context of climate change, the IPCC considers vulnerability as "...the degree to which a system is susceptible to, and unable to cope with, adverse effects of climate change, including climate variability and extremes" (IPCC, 2007). In this regard, vulnerability is a function of the degree of exposure, sensitivity and adaptive capacity of a given natural or social system (Nelson *et al.*, 2010). In the context of global change science, the concepts of adaptation, adaptive capacity, vulnerability, resilience, exposure and sensitivity are interrelated (Smit and Wandel, 2006).

Vulnerability as a concept, for example, is a fundamental tool for unfolding states of susceptibility to harm, powerlessness, and marginality of both physical and social systems, and for guiding normative analysis of actions to enhance well-being through reduction of risk. In its ambit, key parameters of vulnerability are the stresses to which a system is exposed, its sensitivity, and its adaptive capacity (Adger, 2006). Adaptations on the other hand, are expressions of adaptive capacity, and they represent ways of reducing vulnerability. The interaction of environmental and social forces determines exposures and sensitivities, and various social, cultural, political and economic forces shape adaptive capacity. The overlap recognizes that the processes driving exposure, sensitivity and adaptive capacity are frequently interdependent. Vulnerability, its elements of exposure, sensitivity and adaptive capacity, and their determinants are dynamic, varying over time, by type, by stimulus to stimulus, and tend to be place and system specific (Smit and Wandel, 2006). Owing to the fact that the concepts of vulnerability (resilience, exposure and sensitivity) are intertwined with often overlapping meanings and application, the term thus becomes ill defined.

Understanding and unpacking the dimensions of vulnerability as climate change gains a central focus in influencing strategic development discourses, actions of governments, communities and development actors has become critical. Vulnerability defines resource allocation, areas of focus and interventions, as well as the scale of impact and associated costs arising from an extreme climate event. It is also important because climate change and its impacts has gained significant interest with discourse gradually and steadily shifting from an impacts-led perspective to a vulnerability-led perspective (Adger *et al.*, 2004). The vulnerability perspective has gained momentum because it provides an opportunity to examine the underlying socio-economic and institutional factors, and, to a lesser extent, the political and cultural factors that determine how people respond to and cope with climate hazards.

This approach is also vital in assessing people's needs in terms of adaptation and/or improvements needed in their ability to cope with existing threats (Adger *et al.*, 2004). According to Segnestam (2004) and Fellmann (2012), as vulnerability is hinged on the three dimensions of exposure, sensitivity, and adaptive capacity, its analysis often addresses these major dimensions. Owing to the spatial and temporal dimension of vulnerability, production systems tend to exhibit variable dimensions at distinguishable degrees and margins of influence. Studies such as those by Orindi and Murray (2005), Smit and Wandel (2006), Dougill *et al.* (2010), Opiyo *et al.* (2014), and Ayantunde *et al.* (2015) undertaken on the vulnerability of pastoralists in Africa and eastern Africa, in particular, have barely examined the multiple dimensions of vulnerability; yet building the resilience of these communities to climate change and variability depends on a thorough understanding and evidence of such dimensions (Adger, 2006; Vogel *et al.*, 2007; Miller *et al.*, 2010; Dougill *et al.*, 2010; Sendzimir *et al.*, 2011). This chapter examines the dimensions of vulnerability to climate change and variability in pastoral production systems of East Africa with a view of clarifying the degree and extent of exposure, sensitivity and adaptive capacity and their linkages.

Exposure to Climate Change and Variability

The semi-arid and arid lands of East Africa support thousands of pastoralists who herd their livestock, opportunistically exploiting heterogeneous landscape resources (Kimani and Pickard, 1998; Galvin *et al.*, 2004; Little *et al.*, 2014). These areas remain highly susceptible to the vagaries of climate change and variability as it portends a precarious future for the sustainability of livestock production by causing livestock mortality and worsening household food security (Opiyo *et al.*, 2014), raising the fact of their vulnerability to climate change and variability.

The IPCC has defined exposure as "the nature and degree to which a system is exposed to significant climatic variations. Climatic stimuli represents all the elements of climate change, including mean climate characteristics, climate variability, and the frequency and magnitude of extremes" (IPCC, 2001). In that regard, exposure represents the background climate conditions and stimuli against which a system operates (Fellmann, 2012). According to Adger (2006), exposure as a component of vulnerability represents the extent to which a system is subjected to significant climatic variations, as well as the degree and duration of such variations. Temperature increases, variability and change in rainfall among others are climatic variations that can be aggregated as climate variability in the climate system during vulnerability assessments (Fellmann, 2012).

Where systems are exposed to future climate change, the pattern and degree of exposure can be significantly altered. In that case, climate change may be manifested through increased temperature, changing rainfall patterns and weather uncertainties (Ojwang' *et al.*, 2010). In exposure, the perturbations may arise externally and/or internally to the system, interacting with the system to induce a significant transformation within the system, this transformation may often be slow and/or sudden (Gallopin, 2006).

A comprehensive and detailed analysis of project climate changes in Uganda was made by USAID and Tetra Tech ARD (2013). No significant changes in annual rainfall were detected for a 60 year historical period (1951-2010), nor were significant changes predicted for the period 2015-2045. However, a significant decline in rainfall was observed in the arid and semi-arid areas of north-eastern Uganda and south-western Uganda, using historical data, and the decreases occurred mostly in the dry season. Temperatures, on the other hand were observed to have risen by between 0.9-1.5°C over the last 60 years, and were predicted to rise by up to 2°C by 2030. This is expected to increase the risk of diseases and pests in the agricultural sector.

Another observation was that the timing and duration of rainfall can vary very much, with the onset of rains shifting by 15-30 days (earlier or later) and the duration changing by 20-40 days from year to year. According to Mwanthi (2015) rainfall is predicted to be highly variable in the arid lands of the eastern region of East Africa and lowest in the western sector. But in contradiction to USAID and Tetra Tech ARD

(2013), the western part of eastern Africa and much of Uganda are predicted to have more rainfall.

East Africa has various production systems, including agrarian, pastoral and agro-pastoral. From East Africa's perspective, agricultural vulnerability to climate change can be described in terms of exposure to elevated temperatures, the sensitivity of crop and livestock yields to elevated temperatures and the ability of farmers and/or pastoralists to adapt to the effects of this exposure and sensitivity by, for example, planting crop varieties that are more heat-resistant or switching to another type of crop/livestock species (Fellmann, 2012).

Sensitivity to Climate Change and Variability

Sensitivity as a concept has been defined differently by scientists such as Gallopin (2003), Leurs (2005), Adger (2006), Smit and Wandel (2006), and Fellman (2012). According to Gallopin (2003), for example, sensitivity is "...the degree to which the system is modified and/or is affected by an internal or external disturbance or set of disturbances." This can then be measured as the amount of transformation of the system per unit of change in the disturbance (Gallopin, 2006). However, in the context of climate change, the IPCC indicates that the sensitivity of a system to climate change reflects the "degree to which a system is affected, either adversely or beneficially, by climate variability or change. The effect may be direct (*e.g.*, a change in crop yield in response to a change in the mean, range or variability of temperature and precipitation) or indirect (*e.g.*, damages caused by an increase in the frequency of coastal flooding due to sea level rise)" (IPCC, 2007). Accordingly, sensitivity reflects the responsiveness of a system to climatic influences, and the degree to which changes in climate might affect it in its current form (Fellmann, 2012).

The global impact of climate change and variability is evident with the developing nations being hardest hit (Parry *et al.*, 2004; Pen *et al.*, 2004; Labat *et al.*, 2004; Collier *et al.*, 2008; Hertel *et al.*, 2010; Mendelsohn *et al.*, 2012). Annual global damage from extreme weather events increased eightfold between the 1960s and 1990s and between 1980 and 2004, and the cost of global extreme weather events nearly surpassed US 1.4 trillion (Millis, 2005). This pattern has been unceasing since the 1980s with annual global costs ranging from a few US billion dollars to above US$ 200 billion by 2010 (IPCC, 2012). Considerable variation at regional level exists.

Despite this variation at regional level, developing nation states and regions have a higher relative burden arising from extreme climate events of up to 8% of their GDP share (Thornton *et al.*, 2014). By all dimensions, climate change is one of the greatest challenges that Africa faces in the 21st century (Tadesse, 2010). Africa will be severely struck by the impacts of climate change, largely arising from the continent's geographical position and limited adaptive capacity, that is exacerbated by widespread poverty and existing low levels of development (Prospero and Lam, 2003; Collier *et al.*, 2008; Muller *et al.*, 2011).

Africa is likely to suffer from climate change and variability more disproportionately than other regions because it is already hot and dry and the people depend almost entirely on agriculture using outmoded technology and family labour (Ojwang' *et al.*, 2010). Climate change already afflicts Africa and by 2020, it is estimated that between 75 and 250 million people will be exposed to increased water and heat stress due to climate change. In the same vein, crop yields in many countries will considerably be affected, with many expected to experience a 50% decline in yield; this will further constrain Africa's fragile food security situation. Already, global aridity has increased from 17% in the 1950s to 27% in the 2000s and in Africa, between 5 and 8% further increase in arid and semi-arid land is expected by 2080 (Collier *et al.*, 2008).

Climate change already afflicts Africa and by 2020, it is estimated that between 75 and 250 million people will be exposed to increased water and heat stress due to climate change. In the same vein, crop yields in many countries will considerably be affected, with many expected to experience a 50% decline in yield; this will further constrain Africa's fragile food security situation. Already, global aridity has increased from 17% in the 1950s to 27% in the 2000s, and in Africa, between 5 and 8% further increase in arid and semi-arid land is expected by 2080 (Collier *et al.*, 2008).

Pastoralists in East Africa inhabit areas with stochastic climate events making them highly exposed and sensitive to climate change and variability (Verschuren *et al.*, 2000; Luseno *et al.*, 2003; Morton, 2007). Sensitive systems such as these are responsive to climate and are therefore significantly affected by small climate changes (Adger, 2006; Thornton *et al.*, 2009). Climate change and variability are generally expected to have impact on feed crops and grazing systems including among others: changes in herbage growth, changes in the composition of pastures, changes in herbage quality, and greater incidences of drought that may offset any dry matter yield increases (Hopkins and Del Prado, 2007; Thornton *et al.*, 2009).

In pastoral production systems, climate change and variability may manifest as flood and drought events (McPeak and Barret, 2001; Raleigh, 2010; Mongi *et al.*, 2010) and these exposures often get mediated and/ or exacerbated by a marginal political position and weak state capacity (Sabates-Wheeler *et al.*, 2008), which are common features in pastoral areas of East Africa (Lonzy, 2009). According to Schilling and Remling (2011), droughts are becoming more frequent in East Africa. For example, in the period of 1999-2009 Kenya recorded four drought episodes. Rains are now reportedly failing every 2-3 years, instead of the previous 9-10 years.

During the droughts, major losses of livestock are reported (Ojwang' *et al.*, 2010). In the drought of 2009, the Maasai in northern Tanzania lost more than a half of their cattle herd (Goldman and Riosmena, 2013). Another key threat is flooding which sometimes occurs in the same places as where droughts occur. Floods not only kill people and livestock, they also increase the incidence of diseases (Ojwang' *et al.*, 2010; Schilling and Remling, 2011). For example, they may lead to formation of stagnant pools of water that act as breeding grounds for mosquitoes and leading to the spread of malaria. Further, predictions of pastoral livestock production under climate change in Kenya for example show that some 1.8 million cattle could be lost by 2030 because of increased drought frequency, this would translate to a US\$ 630 million loss in terms of livestock value and production foregone (Ericksen *et al.*, 2012). It is also known that the ambient temperature for most domesticated livestock species is between 10 and 30°C and therefore, temperatures above this leads to reduction in feed intake by 3-5% per any additional degree of temperature (Thornton *et al.*, 2014). Already temperature and rainfall changes affect feed quantity and quality in the pastoral areas of East Africa, the projected climate change and variability conditions will only escalate the current patterns.

Predictions further show that climate change will escalate the presence of sudden, extreme and chronic hazards that often destabilize rural, farming and pastoral communities in African countries (Nelson *et al.*, 2010; Xenarios *et al.*, 2014; Alcántara-Ayala *et al.*, 2015). This will further worsen the vulnerability of these communities (Maplecroft, 2014). In the recent past, pastoralists and agro-pastoralists have experienced droughts about one year in five. This increase in drought frequency from one year in five to one year in three sets herd sizes on a rapid and unrecoverable decline path (Thornton *et al.*, 2009).

Adaptive Capacity of East African Pastoral Systems

Adaptive capacity has been referred to in various ways: as a system's coping capacity (Turner *et al.*, 2003), as capacity of response (Gallopin, 2003) and as adaptive capacity (Adger, 2006). Smit and Wandel (2006) show that it is closely related to a host of other commonly used concepts, including among others adaptability, coping ability, management capacity, stability, robustness, flexibility and resilience. Meanwhile, the IPCC defines adaptive capacity as "the ability (or potential) of a system to adjust successfully to climate change (including climate variability and extremes) to: (i) moderate potential damages; (ii) taking advantage of opportunities; and/or (iii) coping with the consequences" (IPCC, 2007).

According to Adger *et al.* (2007) adaptive capacity comprises adjustments both in behaviour and in resources and technologies and is a desirable attribute of a system to reducing vulnerability (Engle, 2011). Several researchers (Behnke et al., 1993; Scoones, 1995; Robinson and Berkes, 2011) have shown pastoralist flexibility, coping capacity, and ability to adapt to stresses and shocks. These are practices that they have honed over years of generations and experiences (Opiyo *et al.*, 2014; Egeru *et al.*, 2014b; Egeru, 2015b).

But, having a set of adaptations as pastoralists, as often extoled, does not amount to having adaptive capacity. This is because the system may be adapted to a given environment and specific range of changes within that environment yet with limited capacity to adapt to new kinds of changes, shocks and stresses or surprises to the system (Armitage and Plummer, 2010). In their work with the pastoral Gabra of northern Kenya, Robinson and Berkes (2011) show that multi-level participation with three key elements; deliberation, inclusivity at multiple levels, and development of a network of institutional arrangements connecting the various levels of governance are critical in enabling participation to build social-ecological resilience and increase adaptive capacity among pastoralists.

East Africa including the greater Horn of Africa has diverse climatic patterns, from equatorial to semi-arid and arid climate (WFP/ICPAC, 2014). Communities living in these differentiated climatic zones adapted their production systems to match the rainfall patterns and other associated factors. Subsequently, East Africa has some of the most complex production systems ranging from the highland perennial farming system, maize-mixed farming system, and cereal-root crop mixed farming system to pastoral and agro-pastoral farming systems. In the semi-arid and arid areas of East Africa is a well-developed and adapted pastoral production system. This production system is based on a maximizing strategy for livestock herding, whilst exploiting the highly variable and heterogeneous landscape resources-water and pasture (Ellis and Swift, 1988; Sato, 1997; Egeru *et al.,* 2014b, Egeru *et al.,* 2015a; Egeru, 2015b).

Pastoral communities in East Africa depend on multi-herd stocking (goats, sheep, camels, donkeys, cattle), with community/tribal grouping preference for various livestock stocking combinations. For example, the Eastern Nilotic Maasai, Samburu, Ariaal, Turkana, Karimojong, Jie peoples, and the Eastern Cushitic Boran and Dasanetch peoples value cattle more than camels. Conversely, the East Cushitic Somali, Rendille and Gabbra peoples value camels more than cattle (Sato, 1997). These tribal preferences for particular livestock species coupled with regional ecological differences produce the different pastoral production systems in East Africa.

Three pastoral production systems can be distinguished in East Africa: transhumant, nomadic and agro-pastoral production systems. The nomadic pastoral system is common in areas with very high aridity and high stochastic rainfall (0-200 mm); about 0-50 growing days per annum, with erratic and long range migrations and camels and goats as lead livestock species (Jahnke and Jahnke, 1982). The Gabbra and Turkana of Kenya exemplify this group (Marhsall, 1990). Pastoral nomads in East Africa depend on their livestock for a pastoral diet of milk, blood and meat. They are also highly obsessed with reverence for their livestock, as they define their cultural identity, marriage rites, compensation and maintaining social relationships through agistment reciprocity (Sato, 1997; Agade, 2010).

The transhumant pastoral production system is perhaps the most predominant in East Africa. It is common in areas with high aridity with rainfall ranging between 200-500 mm; about 50-75 growing days per annum, medium to long-range migration and highly mixed livestock species reared (Jahnke and Jahnke, 1982). They have a well-defined migration pattern that is now being altered by climate change. The pastoral Maasai, Samburu, Borana, Galla, Rendille, Pokot, Sukuma and Somali fall under this category. These pastoral groups may from time to time obtain supplemental food stuffs from neighbouring communities (Marshall, 1990; Coppolillo, 2000).

The agro-pastoral production system is a system that defines pastoral communities in transition. The system occurs in areas with medium to moderate aridity with relatively higher rainfall of 400-600 mm and above. The growing days are relatively high: in the region of 75-90 days per annum. Grains constitute a major food supplement to livestock products. They mainly raise cattle, sheep and goats and are mostly dependent on short-range mobility (Jahnke and Jahnke, 1982). The Bokora, Matheniko, Jie, and Dodoth in Karamoja; the Pokot of both Kenya and Uganda, and the Orma, the Afar, and the Borana of Ethiopia exemplify this category (Oba and Kaitira, 2006; Grade, 2008; Egeru, 2015a).

Meanwhile, two emergent pastoral production systems are slowly developing in East Africa. These are the peri-urban and commercial pastoral production systems. These two new pastoral production systems are coming at the heels of the modernization paradigm that argues that pastoralism under the changing land use systems needs to be intensified. Specifically, commercial pastoralism is a production system that invests in capital accumulation in non-reproductive herds which are then sold when the market environment improves (Oba, 2011). These livestock keepers rely on a two-pronged decision making criteria in a drought event: They can either invest in trucking to ferry their herds to commercial ranches, or they rent grazing reserves during periods of stress (Mahmoud, 2006).

Climate change and variability is making pastoral livelihoods very challenging in arid and semi-arid areas. Both the livestock and crops are sensitive to droughts and floods. And even though pastoralists of East Africa have coped with droughts for many centuries, the recent droughts (for example in Kenya) have been difficult to contend with (Schilling and Remling, 2011). The adaptive capacity of the pastoral communities to shocks, such as climate change and variability, is relatively low. There are few options for adapting. The pastoral communities have few livelihood assets. They have low formal education and those that have are not well educated (Schilling and Remling, 2011). Yet knowledge is an important factor for adaptation. Also, the infrastructure to access alternative sources of livelihoods or to escape calamities is low. For example, the roads in perhaps all pastoral communities of East Africa are not passable during the rainy seasons.

The traditional coping and adaptive strategies have traditionally included expansion of the grazing range. The movement used to take place over a long distance; sometimes as long as 150 km, and lasted many days. Some pastoralists, who could afford to, used trucks to move their animals. But as is explained by Dr Onyango in this book, mobility is no longer easy to undertake because of landscape fragmentation. Besides, migration across national borders exposes pastoralists to harassment.

The Kenyan Maasai sometimes cross the border into Tanzania and are exposed to harassment from the Tanzanian authorities.

Movements within the same country are also not without their challenges. Mobility within national borders can be restricted. For example, the pastoralists are not allowed to access protected areas during droughts. At other times, movement through other pastoralist territories leads to violent conflicts over pastures and water. Additionally, pastoralists practice raiding to take control over grazing areas or to restock their herds if they lost their animals to drought (Schilling et al., 2012). Now it is becoming apparent that raiding is becoming commercialized and has a long chain of actors stretching all the way from business men in urban centres who place orders for livestock through corrupt government officials.

The third most commonly used adaptation strategy has been livelihood diversification. Women are increasingly diversifying into small businesses. These businesses, at least among the Turkana, are supported by the women's control over donkeys (the surest way of mobility within an area where the road infrastructure is very poor) (Schilling and Remling, 2011). Some pastoralists have diversified into sale of souvenirs to tourists. But sale of souvenirs is an erratic and an unreliable source of income. Sales to tourists are few and far between because tourists are not there every day and some areas like Turkana are not favoured by tourists because of perceptions of insecurity.

The fourth and final adaptive strategy has been to migrate to towns and urban centres. Many pastoral people are now to be seen in urban centres working; the males are seen working as security personnel, while the females are seen selling beads and traditional medicine and others begging on the streets (Schilling and Remling, 2011).

Conclusion

In conclusion, drylands of East Africa are highly vulnerable to climate change and vulnerability because of the increased likelihood of droughts and floods and associated incidences of pests and diseases that cause negative impacts that include livestock death and out migrations of people. Additionally, the dryland regions have low adaptive capacity and their adaptive strategies are outdated. For example, transhumance is no longer easy because much of the grazing lands have been fragmented.

In the final analysis, resilience against climate change and variability is a public good to be delivered by governments. Governments may take any of the following actions to build resilience: construct dams, wells and boreholes; and design livestock insurance schemes. Other initiatives include conservation of water catchment areas, improvement of range management to sustain access to pasture, introduction of hay making technologies, transfer of flood waters to drought prone areas, introduction of rain water harvesting technologies and flood control mechanisms, disposal of livestock before onset of drought, and creating small credit schemes (Ojwang' *et al.*, 2010).

References

Adger, W.N., Brooks, N., Bentham, G., Agnew, M., and Eriksen, S. (2004). *New indicators of vulnerability and adaptive capacity* (Vol. 122). Norwich: Tyndall Centre for Climate Change Research.

Adger, W.N., Agrawala, S., Mirza, M.M.Q., Conde, C., O'Brien, K., Pulhin, J., Pulwarty, R., Smit, B. and Takahashi, K. (2007). "Assessment of adaptation practices, options, constraints and capacity." In: *Climate Change 2007: Impacts, adaptation and vulnerability.* Contribution of Working Group II to the Fourth Assessment Report of the Intergovernmental Panel on Climate Change, edited by M.L. Parry, O.F. Canziani, J.P. Palutikof, P.J. van der Linden and C.E. Hanson. Cambridge, UK: Cambridge University Press; pp. 717–743.

Adger, W.N. (2006). "Vulnerability." *Global Environmental Change* 16: 268–281.

Agade, K.M. (2010). "Complexities of livestock raiding in Karamoja." *Nomadic Peoples,* 14(2), 87-105.

Alcántara-Ayala, I., Altan, O., Baker, D., Briceño, S., Cutter, S., Gupta, H., Holloway, A., Ismail-Zadeh, A., Díaz, V.J., Johnston, D., McBean, G., Ogawa, Y., Paton, D., Porio, E., Silbereisen, R., Takeuchi, K., Valsecchi, G., Vogel, C., Wu, G., Zhai, P. (2015). *Disaster risks research and assessment to promote risk reduction and management.* http://www.icsu.org/science-for-policy/disaster-risk/documents/DRRsynthesisPaper_2015.pdf .

Armitage, D., and Plummer, R. (eds.), (2010). *Adaptive capacity and environmental governance.* Berlin: Springer.

Ayantunde, A.A., Turner, M.D., and Kalilou, A. (2015). "Participatory analysis of vulnerability to drought in three agro-pastoral communities in the West African Sahel." *Pastoralism,* 5(1):1-11.

Bassett, T.J., and Fogelman, C. (2013). "Déjà vu or something new? The adaptation concept in the climate change literature." *Geoforum,* 48: 42–53.

Bollig, M. and Schulte, A. (1999). "Environmental change and pastoral perceptions: Degradation and indigenous knowledge in two African pastoral communities." *Human Ecology* 27(3): 493-514.

Collier, P., Conway, G., and Venables, T. (2008). "Climate change and Africa." *Oxford Review of Economic Policy*, 24(2), 337-353.

Coppolillo, P.B. (2000). "The landscape ecology of pastoral herding: spatial analysis of land use and livestock production in East Africa." *Human ecology*, 28(4):527-560.

Dougill, A.J., Fraser, E.D., and Reed, M.S. (2010). "Anticipating vulnerability to climate change in dryland pastoral systems: using dynamic systems models for the Kalahari." *Ecology and Society*, 15(2).

Egeru, A. (2015). "Climate risk management information: Sources and responses in a pastoral region in East Africa." *Climate Risk Management.* (In press).

Egeru, A., Wasonga, O., Mburu, J., Elhadi, Y., Majaliwa, M.G.J., MacOpiyo, L., and Bamutaze, Y. (2015a). "Drivers of forage availability: An integration of remote sensing and traditional ecological knowledge in Karamoja Sub-region, Uganda." *Pastoralism: Research, Policy and Practice*, 5:19.

Egeru, A., Wasonga, O., MacOpiyo, L., Mburu, J., Tabuti, J.R., and Majaliwa, M.G. (2015b). "Piospheric influence on forage species composition and abundance in semi-arid Karamoja sub-region, Uganda." *Pastoralism: Research, Policy and Practice Journal*, 5(1):1-17.

Egeru A., Barasa, B., Massa H.M., and Nampala, P. (2015c). "Piosphere Syndrome and Rangeland Degradation in Karamoja Sub-region, Uganda." *Resources and Environment*, 5(3), 73-89.

Egeru, A., Wasonga, O., Kyagulanyi, J., Majaliwa, M.G.J., MacOpiyo, L., Mburu, J., (2014a). "Spatio-temporal dynamics of forage and land cover changes in Karamoja sub-region, Uganda." *Pastoralism: Research, Policy and Practice Journal*, 4:6.

Egeru, A., Osaliya, R., MacOpiyo, L., Mburu, J., Wasonga, O., Barasa, B., Said, M., Aleper, D., and Majaliwa, M.G.J. (2014b). "Assessing the spatio-temporal climate variability in semi-arid Karamoja sub-region in northeastern Uganda." *International Journal of Environmental Studies*, 71(4): 490-509.

Engle, N.L. (2011). "Adaptive capacity and its assessment." *Global Environmental Change*, 21: 647–656.

Fellmann, T. (2012). "The assessment of climate change-related vulnerability in the agricultural sector: Reviewing conceptual frameworks." In: *Building resilience for adaptation to climate change in the agriculture sector*, pp. 23-37.

Gallopín, G.C. (2006). "Linkages between vulnerability, resilience, and adaptive capacity." *Global Environmental Change*, 16(3): 293-303.

Gallopín, G.C. (2003). "A systemic synthesis of the relations between vulnerability, hazard, exposure and impact, aimed at policy identification." In: *Handbook for estimating the socio-economic and environmental effects of disasters*, by the Economic Commission for Latin American and the Caribbean (ECLAC). Mexico, DF: ECLAC.

Galvin, K.A., Thornton, P.K., Boone, R.B., and Sunderland, J. (2004). "Climate variability and impacts on East African livestock herders: the Maasai of Ngorongoro Conservation Area, Tanzania." *African Journal of Range and Forage Science*, 21(3): 183 189.

Goldman, M.J. and Riosmena, F. (2013). "Adaptive capacity in Tanzanian Maasailand: Changing strategies to cope with drought in fragmented landscapes." *Global Environmental Change*, 23: 588-597.

Gradé, J.T., Tabuti, J.R., and Van Damme, P. (2008). "Ethnoveterinary knowledge in pastoral Karamoja, Uganda." *Journal of Ethnopharmacology*, 122(2): 273-293.

Hertel, T.W., Burke, M.B., and Lobell, D.B. (2010). "The poverty implications of climate-induced crop yield changes by 2030." *Global Environmental Change*, 20(4): 577-585.

Hopkins, A., and Del Prado, A. (2007). "Implications of climate change for grassland in Europe: impacts, adaptations and mitigation options: A review." *Grass and Forage Science*, 62(2): 118-126.

IPCC (2012). "Managing the risks of extreme events and disasters to advance climate change adaptation." In: *A Special Report of Working Groups I and II of the Intergovernmental Panel on Climate Change*, edited by C.B. Field, V. Barros, T.F. Stocker *et al.*, Cambridge, NY: Cambridge University Press.

IPCC (2007). Climate Change 2007: Impacts, adaptation and vulnerability. Summary for policy makers. Available at: www.ipcc.cg/SPM13apr07.pdf (accessed 3 November 2012).

Jahnke, H.E. (1982). *Livestock production systems and livestock development in tropical Africa* (Vol. 35). Kiel: Kieler Wissenschaftsverlag Vauk.

Kimani, K., and Pickard, J. (1998). "Recent trends and implications of group ranch sub-division and fragmentation in Kajiado District, Kenya." *Geographical Journal,* pp. 202-213.

Labat, D., Goddéris, Y., Probst, J.L., and Guyot, J.L. (2004). "Evidence for global runoff increase related to climate warming." *Advances in Water Resources,* 27(6): 631-642.

Little, P.D., Debsu, D.N., and Tiki, W. (2014). "How pastoralists perceive and respond to market opportunities: The case of the Horn of Africa." *Food Policy,* 49: 389-397.

Luers, A.L. 2005. "The surface of vulnerability: an analytical framework for examining environmental change." *Global Environmental Change,* 15: 214–223.

Luseno, W.K., McPeak, J.G., Barrett, C.B., Little, P.D., and Gebru, G. (2003). Assessing the value of climate forecast information for pastoralists: Evidence from Southern Ethiopia and Northern Kenya. *World Development,* 31(9): 1477-1494.

Mahmoud, H.A. (2006). "Innovations in pastoral livestock marketing: the emergence and the role of 'Somali cattle-traders-cum-ranchers' in Kenya." In: *Pastoral livestock marketing in Eastern Africa: Research and policy changes,* edited by J.G. McPeak and P.D. Little. Warwickshire, UK: IT Publication; pp. 129-144.

Maplecroft, V. (2014.) Climate change and environmental risk Atlas 2015. Available from: http://maplecroft.com/portfolio/new-analysis/2014/10/29/climate-change-and-lack-food-security-multiply-risks-conflict-and-civil-unrest-32-countries-maplecroft/.

Marshall, F. (1990). "Origins of specialized pastoral production in East Africa." *American Anthropologist,* 873-894.

McPeak, J.G., and Barrett, C.B. (2001). "Differential risk exposure and stochastic poverty traps among East African pastoralists." *American Journal of Agricultural Economics,* pp. 674-679.

Megersa, B. (2014). Climate change, cattle herd vulnerability and food insecurity: Adaptation through livestock diversification in the Borana pastoral system of Ethiopia. PhD thesis, University of Honheim.

Available from: https://opus.uni-hohenheim.de/volltexte/2014/957/pdf/
Bekele_Thesis_121213_4114.pdf

Mendelsohn, R., Emanuel, K., Chonabayashi, S., and Bakkensen, L. (2012). "The impact of climate change on global tropical cyclone damage." *Nature Climate Change*, 2(3): 205-209.

Miller, F., Osbahr, H., Boyd, E., Thomalla, F., Bharawani, S., Ziervogel, G., Walker, B., Birkmann, J., van der Leeuw, S., Rockström, J., and Hinkel, J. (2010). "Resilience and vulnerability: complementary or conflicting concepts?" *Ecology and Society*, 15(3).

Mongi, H., Majule, A.E., and Lyimo, J.G. (2010). "Vulnerability and adaptation of rain-fed agriculture to climate change and variability in semi-arid Tanzania." *African Journal of Environmental Science and Technology*, 4(6).

Morton, J.F. (2007). "The impact of climate change on smallholder and subsistence agriculture." *Proceedings of the National Academy of Sciences*, 104(50): 19680-19685.

Müller, C., Cramer, W., Hare, W.L., and Lotze-Campen, H. (2011). "Climate change risks for African agriculture." *Proceedings of the National Academy of Sciences*, 108(11): 4313-4315.

Mwanthi, A.M. (2015). Investigating linkages between the subtropical Indian Ocean dipole mode and East African rainfall during October to December. Master of Science Degree in Meteorology, University of Nairobi.

Oba, G. (2011). "Mobility and the sustainability of the pastoral production system in Africa: perspectives of contrasting paradigms." *Future of Pastoralism*.

Oba, G. and Kaitira, L.M. (2006). "Herder knowledge of landscape assessments in arid rangelands in northern Tanzania." *Journal of Arid Environments*, 66(1): 168-186.

O'Brien, K., Eriksen, S., Nygaard, L.P., and Schjolden, A. (2007). "Why different interpretations of vulnerability matter in climate change discourses." *Climate Policy*, 7(1): 73-88.

Ojwang', G.O., Agatsiva, J., Situma, C., 2010. Analysis of Climate Change and Variability Risks in the Smallholder Sector: Case studies of the Laikipia and Narok Districts representing major agro-ecological zones in Kenya. Department of Resource Surveys and Remote Sensing (DRSRS) and FAO, Rome.

Opiyo, F.E., Wasonga, O.V., and Nyangito, M.M. (2014). "Measuring household vulnerability to climate-induced stresses in pastoral rangelands of Kenya: Implications for resilience programming." *Pastoralism,* 4(1): 1-15.

Orindi, V.A., and Murray, L.A. (2005). Adapting to climate change in East Africa: A strategic approach (No. 117). International Institute for Environment and Development. Available from: http://dlc.dlib.indiana. edu/dlc/bitstream/handle/10535/6094/117.pdf?sequence=1&isAllowed=y.

Parry, M.L., Rosenzweig, C., Iglesias, A., Livermore, M., and Fischer, G. (2004). "Effects of climate change on global food production under SRES emissions and socio-economic scenarios." *Global Environmental Change,* 14(1): 53-67.

Peng, S., Huang, J., Sheehy, J.E., Laza, R.C., Visperas, R.M., Zhong, X. and Cassman, K.G. (2004). "Rice yields decline with higher night temperature from global warming." *Proceedings of the National Academy of Sciences of the United States of America,* 101(27): 9971-9975.

Prospero, J.M., and Lamb, P.J. (2003). "African droughts and dust transport to the Caribbean: Climate change implications." *Science,* 302(5647): 1024-1027.

Raleigh, C. (2010). "Political marginalization, climate change, and conflict in African Sahel states." *International Studies Review,* 12(1): 69-86.

Ribot, J.C. (2011). "Vulnerability before adaptation: toward transformative climate action." *Global Environmental Change,* 21: 1160–1162.

Robinson, L.W., and Berkes, F. (2011). "Multi-level participation for building adaptive capacity: Formal agency-community interactions in northern Kenya." *Global Environmental Change,* 21(4): 1185-1194.

Sabates-Wheeler, R., Devereux, S., Mitchell, T., Tanner, T., Davies, M., and Leavy, J. (2008). Rural disaster risk–poverty interface. Report prepared for the Global Assessment Report on Disaster Risk Reduction. Brighton: Institute of Development Studies.

Sato, S. (1997). "How the East African pastoral nomads, especially the Rendille, respond to the encroaching market economy." *African Study Monographs,* 18(3, 4): 121-135

Schilling, J., Opiyo, F.E., and Scheffran, J. (2012). "Raiding pastoral livelihoods: motives and effects of violent conflict in north-western Kenya." *Pastoralism: Research, Policy and Practice,* 2: 25.

Schilling, J. and Remling, E. (2011). Local adaptation and national climate change policy in Kenya: Discrepancies, options, and the way forward. Working Paper, CLISEC-14. Research Group Climate Change and Security. University of Hamburg.

Sendzimir, J., Reij, C.P., and Magnuszewski, P. (2011). "Rebuilding resilience in the Sahel: regreening in the Maradi and Zinder regions of Niger." *Ecology and Society*, 16(3): 1.

Smit, B., and Wandel, J. (2006). "Adaptation, adaptive capacity and vulnerability." *Global Environmental Change*, 16(3): 282-292.

Tadesse, D. (2010). The impact of climate change in Africa. Institute for Security Studies. Available from: https://www.issafrica.org/uploads/Paper220.pdf.

Thornton, P.K., Ericksen, P.J., Herrero, M., and Challinor, A.J. (2014). "Climate variability and vulnerability to climate change: A review." *Global Change Biology*, 20(11): 3313-3328.

Thornton, P.K., Van de Steeg, J., Notenbaert, A., and Herrero, M. (2009). "The impacts of climate change on livestock and livestock systems in developing countries: A review of what we know and what we need to know." *Agricultural Systems*, 101(3): 113-127.

USAID, 2013. Uganda climate change vulnerability assessment Report. USAID & Tetra Tech ARD.

Verschuren, D., Laird, K.R., and Cumming, B.F. (2000). "Rainfall and drought in equatorial East Africa during the past 1,100 years." *Nature*, 403(6768): 410-414.

Vogel, C., Moser, S.C., Kasperson, R.E., and Dabelko, G.D. (2007). "Linking vulnerability, adaptation, and resilience science to practice: Pathways, players, and partnerships." *Global Environmental Change*, 17(3), 349-364.

Xenarios, S., Sarker, G.W., Nemes, A., Nagothu, U.S., Biswas, J.C., and Maniruzzaman, M. (2014). "Socio-ecological vulnerability assessment of flood and saline-prone region in rural Bangladesh." *Bioforsk Report*, 9(24). Available from: http://www.riceclima.com/wp-content/uploads/2012/02/.

CHAPTER 6

Role of Culture and Gender in Shaping Rangeland Management and Adaptation

John R.S. Tabuti, Anthony Egeru and Emmy Jakisa

Department of Environmental Management, Makerere University, Uganda

Introduction

The global climate has changed and the world is now approximately 1°C warmer than it was 100 years ago. In Uganda the temperature rose by 0.9°C between 1950 and 2000. The impacts of climate change are affecting all aspects of human life and will reverse development gains that the world has achieved in the past 200 years. Climate change and variability can manifest as sudden natural disasters of extreme weather events such as flash floods or as slow gradual disasters, *e.g.* floods and drought. In Uganda droughts are becoming more prolonged, wind speeds are increasing, and lightning and thunderstorm are becoming more frequent in the semi-arid district of Nakasongola (Muhanguzi *et al.*, 2012).

In rangelands, climate change and variability affect the health of livestock on which humans depend. For example, the drought of 2009 that took place in northern Tanzania decimated upwards of 55% of the local herds in the Maasai sections of Emanyara and Longido (Goldman and Riosmena, 2013). Climate change and variability also affect the availability and quality of biological natural resources. Furthermore, they affect soil quality through erosion, cause droughts and floods, reduce income generation opportunities for the local populations and even erode basic human rights.

Hence, while climate change and variability invariably affect access to useful natural products and livelihoods, they also affect the cultural domain of rural and natural resource dependent communities (Goldman and Riosemena, 2013). The current cultural alteration, for example, is immense and the turnaround is unceasing. How climate change and variability has affected these communities is mainly observed in how it has altered the gender roles. For instance, among the pastoral communities of East Africa, the household provisioning function for food and daily requirements burden has been shifted to women, after men have lost their livestock herds. Some women are now the principal providers for their households through manual labour. Additionally, girls have been relegated to work as domestic servants in the cities and towns of Uganda. Elsewhere they have been relegated to prostitution and/or street begging.

On the other hand, the male members in pastoral communities are losing their position in society because of the loss of the role of providing security to their community that is tied around livestock keeping. Further, the functional role and position occupied by the elders has been tremendously jeopardized, as their role as guardians of societal values and ethics, peace arbiters and power centres has been lost. For example, among the Karimojong of Uganda, public policy and interventions have reduced cattle raiding and promoted sedentarisation; and these factors have lowered incidences of cattle raiding and with that the need for negotiating powers and skills among the elders has also diminished.

Pastoralists of East Africa

As already pointed out elsewhere in this book, pastoralism is one of the main livelihood options in East Africa and the Horn of Africa (Kipuri and Ridgewell, 2008). There are not many livelihoods that are suited to this arid environment, but mobile livestock keeping is particularly well adapted. Pastoralism is an adaptation to arid and semi-arid climatic conditions in which crop farming cannot be easily sustained due to inadequate or unpredictable rainfall.

Arid and semi-arid areas of the world are characterized by intermittent short rain periods and long drought periods. This type of climate does not favour the crop agriculture that is common to wetter areas. Pastoral communities are masters of managing stochastic resource availability through opportunistic utilization of available rangeland resources, including pasture and water in heterogonous landscapes. These actions of opportunistic utilization have distinguished pastoralists as unique environmental conservationists of fragile environments.

Pastoralists usually have large herds and inhabit large tracts of communally managed land. They make use of heterogeneous landscapes that have patchy distribution of resources, in both time and space, and make the best use of these resources, especially of water and herbage. Vagaries of nature ensure that livestock numbers do not exceed the carrying capacity of the land. For example, pastoralist herds increase in size when conditions are right and are reduced during unfavourable events like drought (Hesse and Cavanna, 2010).

The way of life of pastoralism is characterized by frequent movements within the territory of pastoralists with their livestock. It is because of these migratory practices that pastoralists are adapted to survive in arid and semi-arid areas. Pastoralists use kinship ties for peaceful access to grazing areas (Fratkin, 2001). Rangelands are characterized by "patches" of high ecological value alongside large expanses of marginal range and shrub lands. "Because of the uneven nature of dry landscapes, herders often move their animals to capture variations in forage and water availability. Such landscape variations can occur because of differences in elevation or at a given elevation due to localized rainfall (or both effects combined). The value of these patches is especially revealed during dry seasons and droughts, when herds can be decimated within three to four months" (Little and McPeak, 2014).

In East Africa, pastoralists occupy 70% of the total land of Kenya, 50% of Tanzania and 40% of Uganda (Fratkin, 2001). Pastoralist societies in East Africa include the cattle-keeping and camel keeping groups. Cattle keeping groups include the Maasai in parts of southern Kenya and northern Tanzania; the Turkana of north-western Kenya and their counterparts, the Karamajong, across the border in north-eastern Uganda. Others are the Basongora in western Uganda and the Bahima south-western Uganda; the Samburu, the Borana and the Orma.

Camel keeping pastoralists are mostly found in the drier north-eastern Kenya, southern Ethiopia and Somalia. The Gabra, Rendille and pastoral Somali are the principal camel rearing groups (Fratkin, 2001). A detailed description of the distribution of these pastoralists by group size in East Africa is provided by Fratkin (2001).

Pastoralists comprise the nomadic pastoralists, transhumants, agropastoralists and ranchers. Pastoral nomads follow a seasonal migratory pattern that can vary from year to year. They move around in the rangelands, opportunistically following availability of resources (water and fodder) for their animals. The timing and destination of migrations are determined primarily by the need for water and fodder. An example of this type of pastoralism is seen among the Ngisonyoka in the Turkana region of north-western Kenya who practice a high degree of mobility which has no regular pattern, but depends on the availability of water and pasture (McCabe, 1990). They and other mobile pastoralists in the region (for example, the Maasai) reserve some areas of pasture to be used in critical times (Goldman and Riosmena, 2013).

On the other hand, transhumance is the regular movement of herds between fixed points in order to exploit the seasonal availability of pastures and water (FAO, 2001). A characteristic feature of transhumance is herd splitting; the herders take most of the animals to search for grazing, but leave the resident community with a nucleus of lactating females. Thus transhumants maintain a home base and a satellite herd in search of water and pasture. Transhumance is practiced widely in East Africa, *e.g.* among the Karimojong of Uganda and the Pokot of western Kenya who migrate with their animals across the border into Uganda in search of pasture. The men move with the satellite herds while the women stay back looking after the children and the nucleus herds (Karmebäck *et al.*, 2015).

Agropastoralists are largely sedentary, combining livestock and crop production. Ranchers are pastoralists who own land privately and commonly fence off such land for livestock production. Additional information about these different types of pastoralism is described by Rugadya *(n.d.)*.

Mobility is a vital strategy of the pastoralist for utilizing available forage, by migrating to other regions when grass and water become scarce (Thurow *et al.*, 1989). Migration also helps to evade diseases and pests such as flies and ticks. Mobility enables pastoralists to manage their environment, in that during the process, their animals help disperse seeds which later germinate and increase pasture cover over their pastoral areas (Secretariat of the Convention on Biological Diversity, 2010). The other environmental benefit of mobility is that the distribution of the utilisation of resources by livestock to many different locations allows the used areas to recover. Pastoralists normally use rich patches of vegetation set aside as a reserve for livestock grazing during dry seasons (Little and McPeak, 2014).

Most of the major pastoralist groups move across national boundaries: the Afar between Ethiopia, Eritrea, and Djibouti; Somalis between Ethiopia, Djibouti, Somaliland, Somalia, and Kenya; Borana between Ethiopia and Kenya; and the 'Karamoja cluster' between Kenya, Uganda, and Sudan. These cross-border identities render pastoralists vulnerable in the political cultures of nation states (Oxfam, 2008).

Pastoralists of East Africa (*e.g.* the Gabbra and Karimojong) keep a variety of livestock like cattle, camels, goats, sheep and donkeys (CARE Ethiopia, 2009). This diversification spreads and minimizes risk in pastoral areas (Schilling *et al.*, 2012). In addition, they always keep a large proportion of female animals which is desirable for greater milk production and enhances the potential for rapid increase in herd size. Consequently, females are rarely sold and compose at least 70% of the herd (Thurow *et al.*, 1989).

Culture and Pastoralism

According to the *Advanced English Dictionary* (WordNet® version 3.1), culture is defined as "the attitudes and behaviour that are characteristic of a particular social group or organization". Culture also includes all the knowledge and values shared by a society, and influences how people understand and interpret natural phenomena such as climate change and variability, and also how they respond to changes in their territory. For example, in Burkina Faso the Fulbe, unlike the Rimaiibe, when faced with drought, failed to adopt additional successful livelihood strategies that included labour migration, working for development projects, gardening, and the engagement of women in economic activities, as these were alien interventions in the area (Nielsen and Reenberg, 2010). Knowledge, attitudes and behaviours evolve over time to help the people to adapt to local conditions and enable people to survive in their environment (Adger *et al.*, 2013).

Enormous traditional ecological knowledge important for exploiting marginal resources in rangelands abounds among pastoralists. Such knowledge includes classification of the environment (usually correlated with vegetation and soil type) (Oba, 1985). The Maasai, for example, can identify plant species that fatten cattle and those that increase milk yield (FAO, 2006). Identification and assessment of plant species is usually based on nutritive value, vigour and animal performance. The Maasai can also locate areas with water in their rangelands by observing the behaviour of certain butterfly species. These butterflies usually visit places with high soil moisture, and are a sign of water availability close to the surface. The presence of *Ficus sycamores* (*Orng'aboli*) or other big succulent tree species such as *Acacia ethiopicus* is also an indicator of a high water table (Mung'ong'o, 1995; Saitabau, 2014; Yamat, pers. comm.).

Among the Maasai, cultural knowledge of the environment focuses on use of resources. The Maasai are concerned, for example, with the quality of water and have ways of telling good water from bad. They believe that bad water contains worms that affect their animals' health and consequently leads to reduced animal productivity and fewer animal products. The quality of water is especially an issue during the dry season (FAO, 2006). The Maasai also have a group of diviners locally known as the *Iloibonok*, who can predict calamities and future events and provide remedies. The diviners are consulted for advice during major community calamities, weather changes and sometimes are consulted in matters of diseases (Saitabau, 2014). Furthermore, the Maasai have taboos prohibiting consumption of wild meat. The same culture exists among the Bahima of south-western Uganda.

Among pastoral communities, including the Karimojong and Maasai, indigenous knowledge is used to manage their livestock and livelihoods. Among the Karimojong, for example, traditional knowledge is used to divide livestock into different herds. The herds are separated according to the prevailing resources and health status such as the healthy, the milking, the calves and the sick. The sick herd, some milking cows and the calves remain close to the homestead under the care of women and children, while the healthy herds are grazed in far off pastures (Nalule, 2010). Indigenous knowledge is also used to supplement the food supply by gathering wild fruits and vegetables. Wild plant food, primarily fruit, roots, and honey, provides an important, but understudied contribution to the pastoralists' diet and health.

Cultural institutions are important in shaping adaptation strategies and mediating the adaptive capacities of communities. Cultural institutions include norms, rules, relations and practices (Goldman and Riosmena, 2013). Specifically for the case of East African pastoralists, the cultural institutions such as age sets, kinship ties, elders and diviners have traditionally been important in mediating access to and control of rangeland resources, especially pastures and water.

Some of the cultural institutions that are related to land use management can be exemplified from among the Maasai and the Karamojong. The Maa-speaking people of northern Tanzania and southern Kenya are traditionally divided into sections or tribes (Mung'ong'o, in this book). Each section is managed as a separate common property system, with social rules directing resource management, access and control. Sections are linked to other sections through joint institutional frameworks that are organized around clan and age-set affiliations (McCabe *et al.*, 2010; Goldman and Riosmena, 2013). In the Maasai tradition, pasture was often accessed and managed through informal rules within and between section boundaries. Special reserve pastures were maintained for calves and sick cattle (Goldman and Riosmena, 2013).

Among the Karimojong, the indigenous pastoral resource management systems were also commonly based on distinctive customary rules governing access to and control of resource use. Authority was enshrined among elders. To appropriate the powers, all Karimojong males had to go through a series of five age and two generation sets (the juniors and the elders) (Rugadya, *n.d.*). Specific rules about initiation from one age set to the next and from the generation of the juniors to the elders were enshrined in the community elders. The elders wielded more authority depending on the number of animals that they owned. Elders commonly took the lead in making decisions concerning when and where to migrate and in what formations to do so. Furthermore, they led negotiations on access to communal grazing lands with other sections or tribes (Rugadya, *n.d.*).

In addition to the elders, diviners also contributed to community decision making. The elders not only commanded resources at the household and community levels, they also shaped the behaviour of children and grandchildren (Reda, 2011). Similar age sets are seen among the Oromo. The *Gada* system among the Oromo, for example, organizes males into sets based on age, in which each set has certain responsibilities (Oba, 2009). As discussed above (Rugadya, *n.d.*) and below, climate change has the potential to erode such elders' authority leading to community breakdown and affecting survival in rangelands.

In the East African pastoral communities, there are also cultural laws and regulations that guide the management of various resources. In the Gabra community, for example, there are actions for collective well digging and protection. Wells, land and pasture are managed for the benefit of the whole community (Robinson and Berkes, 2011). Families in pastoral areas also tend to be polygamous (Elfmann, 2005) as having many children provides the much needed labour for pastoralist activity. Labour is needed for herding, security and maintenance of the *boma*.

Pastoral cultures also have customary or traditional institutions that mediate conflict. The *Heer* system among the Oromo refers to a traditional jurisprudence employed to resolve conflicts within and between clans. Formal institutions such as these can be utilized to incite or resolve conflict. Among the Nyangatom of Ethiopia and South Sudan, and the Turkana of northern Kenya, ritual leaders and elders perform blessings and engage in discussions to resolve conflict.

Cultural institutions also prescribe community behaviours. For instance, young men may seek the permission of elders before raiding, and large conflicts usually require the sanction and blessing of ritual leaders before their initiation (Glowacki and Gönc, 2013).

Culture through cultural institutions helps people overcome risk in the harsh range environments. For example, among the Debra in Kenya and the Karimojong in Uganda, a tradition of livestock exchange exists between friends, relatives and agnates. According to this custom, livestock are offered to friends and kin who have been deprived. Relations of reciprocity enhance access to pastures (including reserve pastures) and water resources. Traditionally, during particularly hard times, no needy livestock keeper was denied access to pastures and water. However, with climate change and general social change, there are indications that this tradition is dying (Goldman and Riosmena, 2013).

The other strategy used by pastoralists to survive in their areas is raiding. Although in recent years, livestock raiding has become more violent and destructive, traditionally, pastoral communities would raid for purposes of restocking their herds, especially after periods of drought or outbreaks of diseases (Schilling et al., 2012).

Moreover, pastoral communities have informal and formal channels for sharing information. The Afar of Ethiopia have, for example, a traditional way of information exchange called the *dagu* through which they shared news about important events. When two Afar people meet, they sit down and spend some time discussing the major economic, social and political events that have taken place recently in their localities (Reda, 2011). The Orma of northern Kenya communicated messages about their environment through folklore. The values of grazing land were expressed in folklore and information about the past environment and present were communicated (Oba, 2009). For example, the cattle folklore called the *darma* describes livestock watering, grazing movements and coping with environmental stresses. Each song contains information about how the livestock respond to changing environmental conditions. Folklore has also been widely used among the Karimojong of north-eastern Uganda.

As already noted above, cultural institutions are currently weakening within the pastoral communities, most importantly because of climate change and variability, public policies and changing social values (Rugadya, *n.d.*). And as the cultures weaken, some of the important practices that have previously sustained them are also getting lost. One such change is the loss of reciprocity. The prolonged droughts and famines of 1971-1973, 1982-1984 and 2009 severely altered the cultural systems of pastoral communities both in Kenya and Tanzania. For example, the Boran and Rendille poor pastoralists migrated to urban areas and agricultural productive areas. They shifted their livelihoods to wage labour as paid herders, night watchmen, charcoal burners, crop cultivators and prostitutes. Furthermore, famine led to the emergence of relief towns such as Korr and Kargi in Kenya (Fratkin, 2001). Similar cultural changes associated with intense drought and a changed landscape have been observed among the Maasai by Goldman and Riosmena (2013).

Some old cultural practices are evolving within the context of changing livelihoods, landscapes and institutions. For example, the way migrations are undertaken is changing and some pastoralists are beginning to use motorized transportation for moving sick animals and calves in search of pastures and water. Other practices that are emerging are the purchase of feed or leasing of exclusive rights to pastures. These new practices are dependent on household entitlements, the capacity of households to have cash to buy feed or lease pasture rights, and to access new forms of knowledge and connections to secure new pasture territories that create new opportunities to move in a fragmented landscape. All these were observed during the harsh drought of 2009 in the districts of Monduli and Longido in Tanzania (Goldman and Riosmena, 2013).

Culture, Gender Roles and Responsibilities among Pastoral Societies in East Africa

Culture shapes gender roles, and is important functionally in gender structuring in pastoral societies. Pastoral societies assign roles and accepted norms of behaviour to women that are different from those of men. Gender roles, rights and responsibilities are differentiated at the household and community level, with women assisted by children being the primary providers of basic household needs, most importantly water, food, firewood and wood for construction. Women are also responsible for the care of small livestock like sheep and goats. In some transhumant and agro-pastoral communities where some cultivation goes on, it is the women who do the cultivation. For instance, the Pokot of north-western Kenya cultivate sorghum and millet on small plots (Karmebäck et al., 2015).

Pastoralists have well defined social and gender roles that are useful for their livelihood. According to Fratkin (1997), for example, among the Karimojong of Uganda, children of 5-6 years are herders, while young men at the peak of physical fitness travel long distances to dry season grazing locations, watering herds at hard-to-access water points in the dry season, moving animals to markets through areas that are potentially vulnerable to raiding and crop damage. Married women, on the other hand, have the responsibility of managing buildings and maintenance of temporary houses, milking, fuel and water provision, cooking and taking care of dependents, and monitoring animal health and nutrition status through close contact with lactating cows. Older men have the role of managing political negotiations concerning grazing, shared herding labour, animal sales, and ceremonies, including weddings, initiations and burials. Generally, it is the men who provide security to the household and the community and this exposes them to the risk of death from cattle raids (Fratkin, 1997).

Among the Pokot when the men move with the satellite herds, women stay behind looking after the children and the nucleus herd (Karmebäck *et al.*, 2015). Women spend several hours a day gathering and processing products for basic household needs. For example, Pokot women spend four hours a day gathering firewood (Rugadya, *n.d.*). This is time that is lost to other livelihood and economic activities. Women end up lacking time for leisure and other productive activities. Firewood collection is not only time consuming and laborious, but also exposes women and girls to physical and sexual violence when they leave the safety of their homes. Girls have low enrolment rates in schools and higher rates of dropout from schools as a result of their responsibilities (Abebe, 2014).

Among the Afar, male adults manage the herd and the household. The husband undertakes the herding, milking and selling of animals (Reda, 2011). Most adult men do not normally carry out manual work unless they are poor and do not have grown-up children. Men are responsible for managing all matters pertinent to the household and the community at large. For example, they defend the family and its herd from wild animals and raiders, settle legal disputes and marital problems, and arrange marriages, bride-price, and ritual ceremonies (Reda, 2011).

The Afar wife, on the other hand, fetches water, grinds grains, and prepares food in the house. She also sells small stock and ropes. Afar women are also actively involved in the preparation of temporary shelters for newly married couples.

Among the Afar, children assume a prominent role in herding and related activities. Calves and small stocks are tended in the nearby areas by boys and girls, and sometimes by female adults (Reda, 2011).

In general, besides their contribution to domestic chores (cooking food, fetching water and taking care of children), pastoralist women look after livestock at home, milk livestock, and market animal products such as hides and skins, or even small stock. When men take herds out for grazing and watering, they usually travel long distances and take several days before returning home. In such cases, the women take on the responsibility of heading their households. The young women who sometimes move with men to graze animals do most of the chores that include milking, tethering the animals, washing and milking, while the men are playing or resting. This leaves pastoralist women with little room for rest (Kipuri and Ridgewell, 2008).

Culture influences the social position of men and women. Among pastoralists in East Africa, girls are trained early in their life to help or work with their mothers, who are submissive to their husbands. As the girls grow into adulthood and later on get married, they are introduced into the same system and occupy the same positions as their mothers in the household and community. The stories, sayings, legends and social practices that are embedded within their culture help to reinforce these prescribed roles and social positions (Dupire, 1963).

Generally, significant gender inequalities exist in terms of access to and control over natural resources and the means of production (income, credit, appropriate technologies, and education). In most parts of Africa, women do not normally own land and if they do they control less land than men. This influences access to and control over land-based natural resources like trees. Because of insecure land tenure, women usually access and control poorer quality resources. In addition, women are commonly excluded from participating in decision making about management of natural resources. In most pastoralist societies, women are not supposed to participate in the same meetings with men or are supposed to keep quiet when they attend (Kipuri and Ridgewell, 2008). Often decisions are made about natural resources without any input from women, yet women have an intimate knowledge, relationship and vested interest in the land and environment around them.

The fact that women do not have the powers to decide and plan ahead about their lives or participate in decision making, makes it difficult to cope with the challenges of climate change and renders them vulnerable (Ongoro and Ogara, 2012). Without the power to decide on family resources and finances, women's ability to manage risks and external shocks, such as diversifying livelihoods, is limited. Gender disparities are accentuated by class, caste, ethnicity, health status, and age. Among the Afar pastoralist in Ethiopia, gender inequality is a risk multiplying factor of women's vulnerability to drought (Abebe, 2014.)

Gender and Climate Change

The main impacts of climate change and variability in pastoral areas of East Africa will be reduced availability of key resources for the home (food, water, and wood for cooking and construction) and for livestock (pasture and water). The declines in pastures and water are likely to increase conflict in many pastoral areas. For example, during the 2005/6 drought in Turkana in Kenya, 40 people were killed during a clash between the Turkana and neighbouring Ethiopian pastoralists. There were also reports of fighting between communities seeking to access grazing land and water in the Kenyan pastoral areas between Isiolo and Wajir Districts (Oxfam, 2008). In Uganda, pastoralists are having land and trespass conflicts amongst themselves, between pastoralists and agriculturalists (in, for example, Bulisa, Kasese and some other districts in the cattle corridor), and between pastoralists and protected areas (*e.g.* Queen Elizabeth and Lake Mburo National Parks).

The impacts in agriculture will include a decline in crop yields and harvests that will result in low food availability. This reduction will impact women disproportionately compared to men because of the women's roles of provisioning for the family and associated responsibilities such as tending to the sick and infirm, tending to small ruminants, sick cows and calves. In communities like the Karamoja, it is the women's responsibility to gather wood for house construction and for cooking food, water for home use and for watering livestock kept at the homestead.

In this case, women will be forced to travel longer distances to access these resources. The social disturbances that usually arise because of lost livelihoods associated with livestock and low food availability will increase the vulnerability of women to acquire HIV/AIDS because of early marriages, forced prostitution and sexual violence. In addition, the responsibilities and work burdens of women will increase after men migrate to towns in search of work. Women will assume the responsibilities of leading a household, and this will be on top of their regular chores. Additional tasks for women will include standing in line for food handouts and walking long distances to access the distribution point.

Contribution of Culture and Gender to Climate Change Adaptation Traditional Coping Strategies for Pastoralists

According to Goldman and Riosmena (2013), adaptation against climate change and other shocks among pastoral communities is weakening mostly because of changes in the landscape and cultural institutions. Nevertheless, a number of traditional coping strategies against droughts associated with climate change and variability are practiced in East Africa and are described below and in Fig. 5.1. Some of these are evolving to align themselves with the new realities of the region, including climate change.

Figure. 6.1: Traditional coping strategies against droughts associated with climate change and variability practiced in East Africa

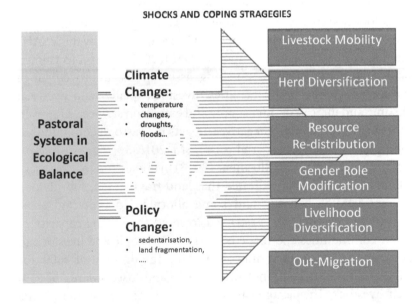

Livestock mobility

The most important and natural traditional coping strategy for pastoralists has traditionally been to move when stresses such as droughts or diseases occur (Bailey, 2012). Migration has traditionally been practiced as risk reducing strategy against disease or drought. Migration in East Africa is becoming difficult to implement in the changed landscape of land fragmentation. Fragmentation disconnects and constrains movement of people and animals to pastures and water.

Land fragmentation is occurring in East Africa through three interrelated processes. One is subdivision of land (dissection) by creating physical, social and administrative boundaries. The second process has been conversion of land, whereby pieces of rangeland are turned into crop land, protected areas or ranches. For example, in south-western Uganda in the early 1990s, communal grazing lands were sub-divided and allocated to a few farmers to create ranches. The last process is compression of land (Karmebäck *et al.*, 2015). This occurs when the extent of pastoral territories is reduced because of increased density of people and livestock.

The fragmentation of land has been promoted by public policy and government measures. In Uganda, for example, many development programmes, such as the Karamoja Livelihoods Programme (KALIP), or the Karamoja Programme, have aimed at improving livelihoods through sedentarisation. In Tanzania, the Range Development and Management Act was followed by the 10-year long Maasai Range Development and Management Project, which commenced in 1970, funded by the USAID and focused on costly technical inputs (UDSM, 2015). The project aimed to achieve the formation of ranching associations; improved range management, animal health facilities, livestock quality and livestock marketing; and increased water supply and extension services. However, the final results were mixed negative impacts on pastoral communities, including the loss of property rights (UDSM, 2015).

A study by Goldman and Riosmena (2013) in the sections of Emanyara and Longido in Tanzania indicated that land fragmented, sub-divided into administrative villages, and some allocated to protected areas where access by people was restricted lowered the coping capacity of the local communities and made movement of people and livestock to escape drought much more difficult. During the drought of 2009, for example, pastoralists could not move freely and lost upwards of 55% of their livestock. The pastoralists who were able to protect their herds were the ones who had money and could purchase feed, lease exclusive rights to grazing areas outside of their territory and move their animals.

Herd diversification

An additional coping strategy is herd diversification. Pastoralists of East Africa (*e.g.* the Gabbra and Karimojong) keep a variety of livestock such as cattle, camels, goats, sheep and donkeys (CARE Ethiopia, 2009). This allows for risk management and spreading in pastoral areas (Schilling *et al.*, 2012). As pastures dwindle, herders increase the number of hardier, browse-dependent goats and camels. They may also adopt species that are more drought resistant or more marketable.

Keeping many different species with different tolerances to environmental stress reduces the risk of losing all livestock to a specific stress (Bati, 2013). Pastoralists also reduce herd sizes during periods of harsh droughts. But this strategy can only benefit the seller when the animals are sold before the droughts have intensified, when the animals are not too thin and before the market is saturated (Goldman and Riosmena, 2013). The pastoralists prefer to sell bulls and to keep the breeding cows (Ongoro and Ogara, 2012).

Livelihood diversification

The second coping strategy that is now commonly practiced by pastoralists is livelihood diversification. The Maasai have diversified their livelihood activities by engaging in wage-labour, eco-tourism, and agriculture, leaving only some communities relying exclusively on livestock (Bailey, 2012). Among the Samburu, some men opt to do away with livestock keeping and migrate to towns to look for employment as guards, while others have turned to brick making and casual manual labour (Ongoro and Ogara, 2012).

Similarly, the Karimojong have diversified their livelihoods to include agriculture, unskilled wage labour, petty trade and local beer brewing. Women are engaged in petty trade activities, including brewing and charcoal burning, as a strategy to cope with environmental shocks that affect agriculture and livestock keeping (Coppock *et al.*, 2006; Ayoo *et al.*, 2012). The male youth are engaged in riding transport motorcycles (*boda boda*). Some youth are now commonly seen in cities like Kampala, Dar es Salaam, *etc.*, selling traditional medicine and cultural trinkets. Casual labour includes working as porters on construction sites, brick-making, ferrying goods for local businessmen, brewing and domestic work (Ayoo *et al.*, 2012). These patterns are observed among all pastoralist communities. For example, the Waso Borana of northern Kenya have diversified into petty trade, wage employment, and farming, as well as the collection and sale of firewood and wild products (Coppock *et al.*, 2006). These activities heavily involve women.

Out-migration is another of the coping mechanisms during stressful years, especially among both the male and female youth. In the struggle towards livelihood diversification some pastoralist youths end up out-migrating permanently from their homelands (Coppock *et al.*, 2006).

Cultural Institutions for Climate Change Adaptation

After all is said and done, it is the cultural institutions that provide the strong social system that helps mitigate risk through the redistribution of resources. Among the Afar of Ethiopia, institutions based on support networks are important for the distribution of livestock and labour resources (Flintan *et al.*, 2008). Re-distributional networks of properties offer support in times of need through kin and in-laws. The paternal kin are responsible for legal protection and compensation, while the maternal kin provide support in times of economic difficulty.

Among the Somali pastoralists a well-developed social system exists which facilitates loans of livestock among families. By distributing the herd, the owner reduces the risk of losses from drought, disease, or aggression. The practice also strengthens goodwill between neighbouring families where by the milk and some of the offspring may be kept by the caretaker (cf. Mung'ong'o (1995) for a description of a similar practice among the Rangi of Central Tanzania). It also increases the prestige of the loaner on the eyes of the community (Thurow *et al.*, 1989).

References

Abebe, M.A. (2014). "Climate change, gender inequality and migration in East Africa." *Washington Journal of Environmental Law and Policy,* 4: 104-140.

Adger, W.N., Barnett, J., Brown, K., Marshall, N. and O'Brien, K. (2013). "Cultural dimensions of climate change impacts and adaptation." *Nature Climate Change,* 3: 112-117.

Ayoo, S., Opio, R. and Kakisa, O.T. (2012). Karamoja situational analysis. Northern Uganda Women's Empowerment Programme (NUWEP) and CARE International in Uganda.

Bailey, J. (2012). Gender dimensions of drought and pastoral mobility among the Maasai. Backgrounder, No. 22, Accessed on the 13.01.16 at 20.10hr from http://dspace.africaportal.org/jspui/bitstream/123456789/32561/1/Backgrounder%20No.%2022.pdf.

Bati, B.M. (2013). Climate change, cattle herd vulnerability and food insecurity: Adaptation through livestock diversification in the Borana pastoral system of Ethiopia. Faculty of Agricultural Sciences. University of Hohenheim.

CARE Ethiopia (2009). Women's milk and small ruminant marketing in Mandera Triangles Kenya, Ethiopia and Somalia. WABEKBON Development Consultants PLC and CARE Ethiopia.

Coppock, D.L., Jillo, A.D. and Aboud, A.A. (2006). From herd diversification to livelihood diversification as a response to poverty: The case of the Waso Boran of northern Kenya. Research Brief 06-05-PARIMA. Global Livestock Collaborative Research Support Program. University of California, Davis. 4 pp.

Dupire, M. (1963). "The position of women in a pastoral society (the Fulani WoDaaBe nomads of the Niger)". In: *Women of tropical Africa*, edited by D. Paulme. Berkeley, California: University of California Press; pp. 47-92.

FAO (2006). The role of indigenous knowledge in range management and forage plants for improving livestock productivity and food security in the Maasai communities. A Links Project Case Study No. 8.

FAO (2001). Pastoralism in the new millennium. FAO Animal Production and Health Paper #150.

Flintan, F., Demlie, S., Awol, M., Humed, Z., Belete, Y. and Lemma, H. (2008). Study on women's property rights in Afar and Oromiya Regions, Ethiopia. USAID and CARE.

Fratkin, E. (2001). "East African pastoralism in transition: Maasai, Boran, and Rendille cases." *African Studies Review*, 44: 1-25.

Glowacki, L. and Gönc, K. (2013). "Customary institutions and traditions in pastoralist societies: Neglected potential for conflict resolution." *Conflict Trends*, 2013, 26 -32.

Goldman, M.J. and Riosmena, F. (2013). "Adaptive capacity in Tanzanian Maasailand: Changing strategies to cope with drought in fragmented landscapes." *Global Environmental Change*, 23: 588-597.

Hesse, C. and Cavanna, S. (2010). Modern and mobile: The future of livestock production in Africa's dry lands. London: International Institute for Environment & Development (IIED) and SOS Sahel International UK.

Karmebäck, V.N., Ndung'u Wairore, J., Jirström, M., and Nyberg, G. (2015). "Assessing gender roles in a changing landscape: diversified agro-pastoralism in drylands of West Pokot, Kenya." *Pastoralism: Research, Policy and Practice*, 5:21.

Kipuri, N. and Ridgewell, A. (2008). A double bind: The exclusion of pastoralist women in the East and Horn of Africa. Minority Rights Group International (MRG).

McCabe, J.T. (1990). "Turkana pastoralism: A case against the Tragedy of the Commons." *Human Ecology,* 18: 81-103.

McCabe, J.T., Leslie, P.W. and DeLuca, L. (2010). "Adopting cultivation to remain pastoralists: The diversification of Maasai livelihoods in Northern Tanzania." *Human Ecology,* 38:321–334.

Muhanguzi, F.K., Kabonesa, C. and Muhanguzi, H.R.D. (2012). Gender and climate change: Assessing impacts and strategies for mitigation and adaptation to climate change in Uganda. School of Women and Gender Studies, Makerere University, and Ministry of Water and Environment, Climate Change Unit.

Mung'ong'o, C.G. (1995). Social processes and ecology in the Kondoa Irangi Hills, Central Tanzania. Meddelanden Series B 93, Department of Human Geography, Stockholm University.

Nalule, A.S. (2010). Social management of rangelands and settlement in Karamoja. Rome: FAO & EU.

Nielsen, J.Ø. and Reenberg, A. (2010). "Cultural barriers to climate change adaptation: A case study from Northern Burkina Faso." *Global Environmental Change,* 20.

Oba, G. (2009). Harnessing pastoralists' indigenous range management knowledge for drought-resilient livelihood systems in the Horn of Africa. WISP, IUCN and FAO.

Oba, G. (1985). "Perception of environment among Kenyan pastoralists: Implications for development." *Nomadic Peoples,* 19: 33-57.

Ongoro, E.B. and Ogara, W. (2012). "Impact of climate change and gender roles in community adaptation: A case study of pastoralists in Samburu East District, Kenya." *International Journal of Biodiversity and Conservation,* 42: 78-89.

Oxfam (2008). Survival of the fittest: Pastoralism and climate change in East Africa. Oxfam Briefing Paper 116.

Reda, K.T. (2011). "Social organization and cultural institutions of the Afar of Northern Ethiopia." *International Journal of Sociology and Anthropology,* 3: 423-429.

Robinson, L.W. and Berkes, F. (2011). "Multi-level participation for building adaptive capacity: Formal agency-community interactions in northern Kenya." *Global Environmental Change*, 21(4): 1185-1194.

Rugadya, M.A. (*n.d.*) Pastoralism as a conservation strategy: Uganda Country Paper. http://cmsdata.iucn.org/downloads/uganda_country_study.pdf.

Saitabau, H.O. (2014). Impacts of climate change on the livelihoods of Loita Maasai pastoral community and related indigenous knowledge on adaptation and mitigation. National Museums of Kenya.

Schilling, J., Opiyo, F.E. and Scheffran, J. (2012). "Raiding pastoral livelihoods: Motives and effects of violent conflict in north-western Kenya." *Pastoralism: Research, Policy and Practice*, 2: 25.

Secretariat of the Convention on Biological Diversity (2010). Pastoralism, nature conservation and development: A good practice guide. Montreal: Secretariat of the Convention on Biological Diversity.

Thurow, T.L., Herlocker, D.J. and Elmi, A.A. (1989). "Development projects and Somali pastoralism." *Rangelands*, 11: 35-39.

University of Dar es Salaam, Centre for Climate Change Studies (2015). Tanzania: Country situation assessment. Report prepared by the Tanzania Pathways to Resilience in Semi-arid Economies (PRISE), Research for Climate-Resilient Futures Team at the Centre for Climate Change Studies (CCCS), University of Dar es Salaam.

PART III

Pastoral Knowledge Systems and
Resilience to Climate Change

CHAPTER 7

The Technical and Social Knowledge Systems of Pastoral Mobility in East Africa

Emma T. Liwenga

Centre for Climate Change Studies, University of Dar es Salaam, Tanzania

Introduction

East Africa is home for thousands of pastoralists who herd their livestock in the arid and semi-arid areas (ASALs) of the region. The indigenous groups of pastoral people living in these semi-arid and arid areas have developed special natural resource management systems. They have developed various strategies to cope with various stress factors affecting their livelihood systems and ability to sustain production.

These strategies, according to Galvin *et al.* (2003), are "important for their own livelihoods but also for the national economies," in terms of contributing "a large share of livestock to markets in the region" as well as coping with climate variability. The key strategies include mobility patterns entailing "moving livestock according to vegetation needs and water availability, keeping species-specific herds to take advantage of the heterogeneous nature of the environment…diversifying their livelihood strategies and, in some cases, temporary or permanent emigration".

However, natural resources in the ASALs are increasingly becoming threatened by climate change and variability. So much so that some of the strategies to cope with climate variability and change are currently not working well. There are several reasons for this phenomenon. The nature of the climate variability that pastoralists are used to dealing with is itself changing (IPCC, 2001). According to Calvosa *et al.* (2010), the repercussions of climate change on the livestock sector are becoming manifest in changes in the quality and quantity of vegetation, availability of fodder and water, and an increase in climate-related diseases.

The current constraints to these traditional strategies have made pastoralists more vulnerable to natural and human-derived perturbations. It is therefore argued that pastoralists no longer can depend on their livestock as the sole basis of their livelihoods; yet they have few opportunities for livelihood diversification (FAO, 2009).

It is argued, therefore, that the traditional ways of pastoralists will continue to change in the context of their mobility patterns, resulting in livelihood diversification to other income-generating activities such as crop production and wage labour. All these are likely to occur while other family members continue to herd the family stock and move to follow the availability of forage. It is thus important to understand the extent to which the pastoral knowledge system contributes to resilience to climate variability and change among other stress factors. Except for Gaiballah and Abdalla (in this book), little has been documented with respect to pastoral mobility patterns, the resilience of pastoralists who practice mobility and climate change.

This chapter focuses particularly on the technical and social knowledge systems of pastoral mobility. The chapter contains critical reviews of the status of knowledge in this thematic area, clearly identifying research and policy issues that need to be addressed in the context of climate smart pastoralism. The chapter further articulates the global, regional and local relevance of the issues raised.

Based on a review of various case studies from within the African region, particularly within the East African region, this chapter analyses how various actors perceive the extent to which the pastoral knowledge system contributes to resilience to climate variability and change. One knowledge system is a scientific one, the "new rangeland paradigm". According to this paradigm, pastoral mobility is a means to balance variability in dryland resources; hence, 'nature' is the point of departure. Another knowledge system is local pastoral knowledge. It is argued that to the pastoralists, the well-being of their animals is the point of departure and mobility is used to ensure that the stock is in good condition.

Adriansen (2005) explains that "…while the positive perception of mobility is relatively new among researchers of drylands, this is not a new line of thought among social scientists studying pastoralists (*e.g.* Dyson-Hudson and Dyson-Hudson 1980; Stenning, 1957). Especially anthropologists … [such as Dyson-Hudson (1966)] and others studying pastoral societies [such as Johnson (1969)]… have pointed to the flexible strategies employed by pastoralists", and which are associated with their mobility patterns.

Furthermore, this chapter shows the extent to which it is important to understand the mobility patterns of pastoralists and of their herd, focusing on their knowledge systems. It is noted, for example, that the pastoralists are willing to move around within a small territory, which they consider their place, but are unwilling to employ large-scale mobility themselves. By looking at the pastoral livelihood systems, perceptions and paradigms explaining their mobility patterns, we achieve a better understanding of how the pastoral knowledge systems contribute to their resilience in the context of climate change.

Pastoralism as a Livelihood System

Understanding the system

Many efforts have been made to define pastoralism, but those who have struggled to do so have failed to embrace the great diversity of pastoral systems that exist around the world, in general, and within Africa, specifically. Some definitions have focused on ethnic identity, whilst others have focused on reliance on livestock (*e.g.* Ellis and Swift, 1988), but none has provided an adequate means of differentiating among various livestock production sub-systems. This may be, in part, due to people having different reasons for categorizing livestock keepers and is common to include additional criteria, such as the use of common property and private resources, the practice of mobility, the use of family labour, husbandry of a mixture of indigenous and cross-bred livestock, *etc.*

It is thus that we have definitions like "...(p)astoralism is the finely honed symbiotic relationship between local ecology, domesticated livestock and people in resource-scarce, climatically marginal and highly variable conditions...a complex form of natural resource management, involving a continuous ecological balance between pastures, livestock and people" (WISP, 2007); or pastoralism is an "...extensive livestock production in the rangelands" (FAO, 2006). The concept **pastoralist** has been used to indicate both a cultural identity and a production or livelihood system. It can refer to people who practice pastoralism, those who share a pastoralist background, or those who are involved in activities related to pastoralism.

It is generally acknowledged, however, that pastoralists are those whose way of life largely depends on mobile livestock herding (OXFAM, 2008). Pastoralists live in a range of environments in many countries across the world. In sub-Saharan Africa, mobile pastoralism is predominantly practised in arid and semi-arid lands (ASALs). The ASALs are areas which experience a hot and dry climate, with low and erratic rainfall. Generally, there are not many livelihoods that are suited to this arid environment, except for mobile livestock keeping which is particularly well adapted.

Pastoralism in Africa evolved in response to climate variability over 6,000 years ago when the Sahara entered a period of prolonged desiccation (OXFAM, 2008). With no reliable supplies of permanent water, pastoralism enabled people to adapt to an increasingly arid and unpredictable environment by moving livestock according to the shifting availability of water and pasture (Brooks, 2006). Apparently, this opportunistic management system continues to this day, making pastoralism an effective and efficient land-use and production system for the drylands of the world.

In order to cope with the varying rainfall and forage distribution, both nomads and their animals must possess a high degree of mobility. According to Oba and Lusigi (1987), two aspects of mobility need to be recognized, *i.e. Resource exploitation mobility* and *Escape mobility. Resource exploitation mobility* is undertaken in response to unpredictable forage and water availability, while *Escape mobility* involves long distance migration to escape drought conditions. In either case, the primary objective is usually to maximise livestock survival.

Generally, there is no single agreed system for classifying livestock enterprises. However, in an effort to ensure appropriate coverage of different forms of livestock production around Africa, it is necessary to be aware of the range of production systems that are in question. Apparently, there is considerable diversity of production systems across the continent, shaped largely by climate (rainfall amount and distribution, as well as temperature) and landscape as well as socio-cultural and economic factors. Figure 6.1 presents the three pillars of pastoralism, integrating the economic, institutional and ecological aspects related to livestock keeping among pastoral communities.

Nevertheless, a number of researchers have shown that pastoralists are not interested in land *per se*. They are only interested in the relationship between human beings, animals and land. In this context, pastoral mobility is important because it is a means to ensure the survival of the herd and, indirectly, of the humans; not because it is good range management (Spencer, 1998; Bollig and Schulte, 1999; Adriansen, 2002; Mung'ong'o, in this book).

Figure 7.1: Three pillars of pastoralism (Source: IIED Evaluation

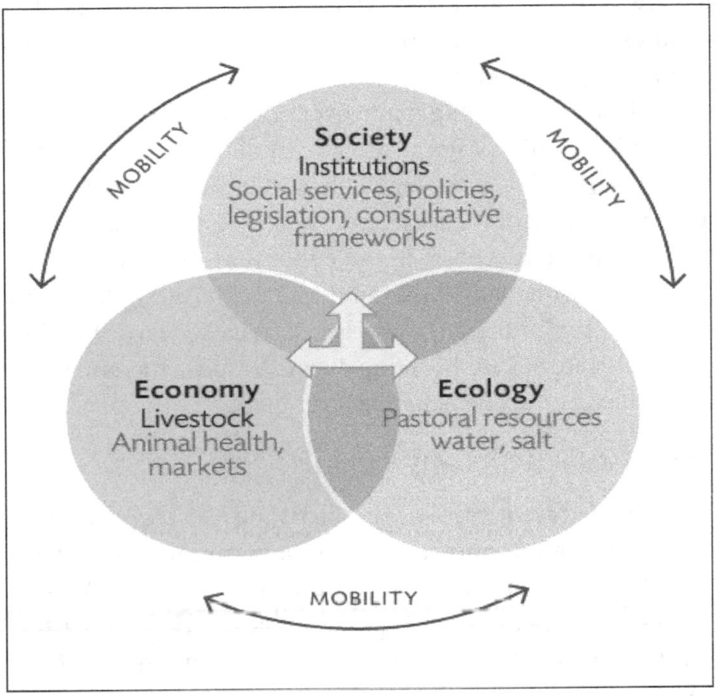

(2013), cited in AFC (2014)).

FAO (2006) categorizes livestock production systems in four main systems, as follows:

1. **Pastoralism** – traditional extensive systems in which availability of grazing dictates herd movements. There are two types of this category:

    a) *Nomadism* involves random movements with the herder's family, during which no permanent settlements are created as people live in temporary shelters; and

    b) *Transhumance* involves seasonal movements along precise routes, practiced in order to exploit the seasonal availability of pastures.

2. **Ranching** - can either be private or state owned in which meat production or breed development is carried out on an enclosed area of land;

3. **Agro-pastoralism** in which livestock are reared extensively and crops are also produced; and

4. **Off-land systems** in which cultivated fodder and agro-industrial by-products are used as animal feed; typically found on the outskirts of towns and expand with urbanization; focusing mainly on poultry, piggery and milk production.

Apparently, drought is not a new phenomenon to the pastoral nomads, since historically pastoralists have suffered numerous such disasters. Their survival can be attributed to a wide spectrum of adaptive strategies. Some of the strategies are ecologically based, while others depend upon socio-economic and cultural mechanisms learned through experience with nature (Mapinduzi *et al.*, 2003). According to Oba and Lusigi (1987), there is, however, serious concern that these drought response strategies are breaking down. As such, currently more pastoralists are seeking outside help than ever before.

Perceptions of Pastoral Mobility Patterns

Pastoralism is regarded as a complex, specialised and effective production system whose three main elements (ecological, economic and societal) work together through strategic mobility. According to AFC (2014), "… (m)obility is often regarded as an important characteristic of pastoral societies and their ways of production in Africa…. the interpretation of the importance of pastoral mobility changes along with the various discourses, and – just as important – the interpretation varies with the professional background of the researcher."

In the context of East African rangelands, Galvin *et al.*, (2003), quoting Coughenour (1991), have characterized them as "… generally heterogeneous, due to spatial gradients of climate, soils, landscape and disturbance…. (t)emporal and spatial variability in rainfall, in particular, increases heterogeneity of the system." As a result of this temporal and spatial heterogeneity, pastoralists keep on exploiting "…spatially distinct areas of vegetation type and productivity by moving species-specific livestock across the landscape." This, according to Galvin *et al.* (2003), implies further that "…livestock and the people who depend on them require access to a complex set of vegetation communities to maintain animal condition, productivity, and population stability, especially under conditions of high climatic variability and other sources of perturbation…."

According to the IUCN, "…(p)astoral mobility is highly diverse driven by both typical and atypical climatic events and seasonal patterns, and by social, economic, political and cultural factors" (IUCN, 2011). Apparently, there are various patterns of mobility. For example, "…

mobility among clearly defined resource areas may be highly regular, sometimes even following clearly defined transhumant routes that have not changed for centuries, or it can be nearly random.... (t)he former is called transhumance and the latter tends to be referred to as nomadism, particularly where livestock keepers do not own defined areas of land" (IUCN, 2011).

In the late 1980s, a new understanding of the drylands dynamic gained importance and led to the so called "new rangeland paradigm" which has been called the "state and transition paradigm (Westoby et al., 1989) or "instability but persistence" paradigm (Warren, 1995). It is apparent that there are different ways of analysing pastoralist mobility patterns, from the rangeland ecological point of view and from the pastoral perspective. Furthermore, it is worth noting that the "new rangeland paradigm" has been thoroughly analysed in the three seminal books Behnke et al. (1993), Scoones (1995a), and Niamir-Fuller (1999a). While the first book is mainly concerned with the ecological aspects of dryland ecosystems, the second one elaborates on the management implications for pastoral production systems, and the last one emphasizes one aspect of pastoral systems, namely pastoral mobility.

According to the new rangeland paradigm, drylands are considered disequilibrial (changing from one state to another) due to strong external controls, e.g. droughts, fires, or insect attacks. These external controls strongly affect primary production and thus livestock density. Hence, dryland productivity is controlled mainly by the highly variable precipitation; because livestock seldom reaches densities high enough to influence vegetation productivity, precipitation is the principal factor controlling inter-annual vegetation dynamics (Coppock, 1993).

In the book on pastoral mobility by Niamir-Fuller (1999a), the so-called "mobility paradigm" is developed. When analysed in the light of aspects such as disequilibrium ecology, management practices and institutions that previously have been characterised as destructive, are now seen as ecologically rational. Hence, the "mobility paradigm" provides a framework for understanding pastoral mobility based on the findings of the new rangeland paradigm and examines the various aspects of mobility.

The perception of pastoral mobility within the new rangeland paradigm relates to the fact that pastoral mobility is highly appropriate in variable and unpredictable environments. An important characteristic of tropical drylands is the heterogeneity of natural resources. Pastoral mobility implies that pastoralists can move to areas with pasture for their livestock. Moreover, pastoral mobility means that the effect of unforeseen events, *e.g.* outbreak of disease, bush fire, locust attack, can be mitigated. Finally, migration between different agro-ecological zones means that more animals can be kept than in each of the zones (Niamir-Fuller, 1998, Scoones, 1995b).

Niamir-Fuller and Turner (1999) have further developed an analytical framework that acts as a checklist for understanding pastoral mobility. The authors aim to ensure appropriate measures that can allow "self-evolution of pastoralism towards an economically, socially and environmentally sustainable livelihood system" (*ibid.*). They find the following four aspects salient for understanding mobility: the resource base, the resource users, their adaptive strategies, and their common property regimes (Niamir-Fuller and Turner, 1999).

The key words for the resource base are high variability and uncertainty, non-equilibrium theory, ecological resilience, and socio-ecological pasture units. Hence, these are the elements of the "new rangeland paradigm" as reflected earlier by Niamir-Fuller and Turner (1999), to provide an understanding of mobility by pastoralists as related to rangeland management. The key words for the resource users are heterogeneity, indigenous technical knowledge system, social capital, reciprocity, interdependence, and political alliance. It is accordingly argued that the complexities of communal resource management and transhumance, pastoral systems necessitate a demand for a high degree of social organization and control. As such, the indigenous technical knowledge and capacity to self-govern is pivotal to the resilience of pastoral communities and many of the challenges that African pastoral societies face today.

The adaptive strategies of pastoralists as among the resource users concern the types of mobility as well as the decision-making and management aspects of mobility. The key words include: opportunistic mobility, tracking, micro-mobility, macro-mobility, negotiation, indigenous communication, and safety nets. To sum up, with the "new rangeland paradigm" and a comprehensive understanding of drylands, it can be seen that mobility is an appropriate strategy in dryland environments.

The Role of Pastoral Knowledge Systems in Mobility Patterns

It is argued that nomads often occupy specific tribal territories (Oba and Lusigi, 1987). Accordingly, grazing lands within a tribal territory are often partitioned into wet season and dry season ranges, whereas wet season ranges are the product of seasonal rains. On the one hand, the wet season ranges areas are dominated by annual vegetation and, except for a short period of the year, remain waterless and uninhabited. On the other hand, the dry season ranges are high potential areas and include pockets of highlands, riparian ecosystems, flood plains or lake shores and the tsetse infested ranges, traditionally considered as drought grazing reserves.

Furthermore, Oba and Lusigi (1987) report that drought strategies used by nomadic pastoralists are highly variable. The two authors further explain that the strategies employed by pastoralists depend upon past drought history, the kinds of livestock raised, and available resources. It is further argued that the entire concept of nomadism may be considered as a means of coping with and exploiting highly variable resources (*ibid.*). This is accordingly made possible in part through the ability of nomads to maintain several species of diverse herds of livestock - camels, cattle, sheep and goats - and by their geographical mobility.

The numbers of animals that pastoralists keep is also considered as a survival strategy. Hidore and Elton (1975), for instance, speculate that the Maasai of Kenya and Tanzania own large herds because these are considered as insurance against drought losses. It is often argued that nomads keep too many animals, far above what the range resources are capable of supporting. The large herds are then unable to survive drought induced periods of low forage production. Nomads, on the other hand, believe large numbers of stock are essential to get through the drought years (Brokensha *et al.*, 1977). This is based upon the theory that the more stock a man has, the better the chance that more of his herd will survive a drought. Eckholm (1975) further argues that in the nomadic family, being largely isolated from the cash economy, more animals are needed to meet basic food requirements. Surplus stock beyond a basic 'minimum' are both an investment and a form of insurance for the drought years when some animals may be sold to buy food grains.

The objective of herd maximisation is best realised through herd diversification (Oba and Lusigi, 1987). Several types of animals are maintained to secure survival of pastoral households. The use of different livestock species has ecological and economic implications. Different species fill different ecological niches and therefore may be more efficient than a single species. Each kind of stock prefers to graze certain plant species and certain types of topography. For example, camels and goats are considered to be browsers, although they may be grazers at certain times. Cattle and sheep, however, are largely grazers. An area containing both grasses and shrubs may be utilised best with a combination of different ruminant species employing different grazing habits (Franke *et al.,* 1980). Different ruminant species display differential digestibility of various dietary constituents. For example, goats have better ability for utilising fodder than do either cattle or sheep (Lundholm, 1976).

During periods of severe drought when forage supplies are critically limiting, livestock survival becomes a major consideration (Oba and Lusigi, 1987). The pastoral nomads of northern Kenya (Borana and Gabra), for example, respond to this problem by moving all livestock from degraded rangelands closer to permanent water, into areas of the range, which because of lack of water, had remained unused (*ibid.*). This strategy does, however, raise some problems for the pastoralists and their livestock. Livestock are kept away from water for a much longer period than during a normal dry season, and distance to water is substantially increased for all livestock species.

Pastoralists need large herds not only to meet household requirements, but also as a means of building social alliances through transfer of livestock to friends and kin as loans. These transactions could then be called upon in times of need (Oba and Lusigi, 1987). Pastoralists manage rangelands through a complex system of common and individual property rights and, where their land has become degraded, it is usually a combination of the weakening of customary management institutions and the loss of key resources that make up the pastoral system (Scoones 1995; Niamir-Fuller 1999). Both of these factors are closely linked with insecurity of land tenure, which is perhaps the single most important challenge facing African pastoral producers. In the context of such challenges, local knowledge systems become imperative in coordinating mobility patterns.

It should be noted that local knowledge systems such as the use of Indigenous Knowledge (IK) plays a significant role for communities in a particular locality to earn their livelihood. IK informs decision making about fundamental aspects of day today life. It is locally bound; indigenous to a specific area; culture and context specific; non-formal knowledge; orally transmitted and generally not documented; dynamic and adaptive; and holistic in nature and closely related to survival and subsistence for many people (Little *et al.,* 2001).

However, pastoral systems in Africa are generally undergoing major social transformations with many commentators suggesting that pastoralism is in decline (IUCN, 2011). Accordingly, there is an indication that for many pastoral societies increasing population density, loss of key pastoral resources and almost static livestock populations all point to the growing pressure on the pastoral system. Communities that used to be considered as nomadic are now less mobile and more likely to have settled homesteads, which brings both benefits in terms of access to services as well as challenges such as natural resource degradation and reduced economic potential (Davies *et al.*, 2010). It is further reported that less than 15% of pastoralists in West and Central Africa are now classified as nomads, whilst in East Africa the figure is probably lower (*ibid.*). The Turkana of northern Kenya are still referred to by some observers as nomads, yet many Turkana now grow crops and have at least one foot in one of the many small towns that have grown up around food aid and development projects in their territory (Majok and Schwabe, 1996).

Loss of key resource tracts and access routes have contributed to the decline in pastoral mobility, whilst at the same time pressure over resources from neighbouring competitors has led many pastoralists to settle simply to protect their own land assets (IUCN, 2010). It is accordingly argued that communities that used to migrate *en masse* with their herds now may remain divided throughout much of the year, often with women, children and some of the elders remaining in the homestead, whilst younger men herd the livestock. For example, the Maasai of Kenya traditionally followed a transhumant migratory pattern returning to a home *manyatta* during the wet season (Majok and Schwabe, 1996). However, 100 years of colonial, governmental and developmental influences have been directed at settling them and introducing the concept of private land ownership (Blewett, 1995). As a result, many Maasai have adopted permanent agricultural and ranching practices, while some of their best dry season pastures have been claimed by non-Maasai communities. Swift (2000) furthermore points out that "commercialisation may well be of outmost importance for future pastoralists...." Commercialisation influences other aspects of pastoral strategies such as safety nets and risk spreading, which have a tendency to disappear under increased commercialisation (see also Batterbury and Warren, 2001; Sutter, 1987; Swallow, 1994).

Among the challenges encountered by pastoral communities in the context of their mobility patterns are the trans-boundary issues. International boundaries created by colonial powers and preserved by independent African countries, have resulted in barring indigenous livestock herds from traditional grazing areas. However, experience shows that during 'regional droughts' nomads disregard these boundaries

(Breman *et al.*, 1978; Gilles and Jamtgaard, 1982). The timing of these long distance migrations is extremely important (Berkes *et al.*, 2000). For example, Kenyan Borana sustained high livestock losses (60-70%) before reaching Southern Ethiopia following the 1971-1973 drought because they waited too long (Legesse, 1980 cited in Oba and Lusigi, 1987). It is apparent that long distance migration across international borders when undertaken on inadequate information can have serious consequences for the pastoralists. A classic example is when the group, hoping to find water along its route, attempts the long journey only to find on arrival that the water source has dried up, or the borehole engine has broken down or even has been vandalised. This implies that various pastoral kinds of knowledge systems and structural organizations form an important integral element of knowledge systems in their mobility patterns.

Conclusion

As noted by Adriansen (2005), the "mobility paradigm" is crucial in understanding and dealing with contemporary pastoral mobility. The paradigm clearly shows that mobility is necessary or rational for pastoralists in arid and semi-arid lands. It shows, as Adriansen correctly concludes, that "in order to fully understand pastoral mobility, its rationale and consequences, it is therefore necessary to address both the range ecology...and the pastoralists' perceptions". Economically logical and environmentally essential, mobility is, in fact, the only way to make sustainable use of rangelands (IFAD, 2009).

Accordingly, the pastoral system is moved to fit the environment in order to make the best use of the available resources. Mobility enables pastoralists to take advantage of pasture resources that are only seasonally accessible, and allows access to salt patches critical for animal health, and other resources and services (*ibid.*). Moreover, thanks to mobility, pastoralists can obtain sufficient supplies of food, forage and water, and avoid disease outbreaks.

Under continuous population change and probable policy changes, resilience is likely to decline. Thus, to really assess vulnerability, we need a comprehensive understanding of complex interactions among climate, ecological, demographic, political and economic systems. This means that pastoralist actions and mobility practices can better be explained using both scientific understanding of drylands and pastoral conceptualisations of nature (Berkes *et al.*, 2000). The absence of pastoralist perspectives means that the importance of mobility for the wider social and cultural life is left unresolved.

References

Adriansen, H.K. (2005). "Pastoral Mobility: A Review." *Nomadic Peoples,* ns volume 9, issues 1& 2, pp. 207-214.

Adriansen, H.K. (2002). "A Fulani without cattle is like a woman without jewellery." *Geographica Hafniensia,* A11, Copenhagen.

AFC (2014). Securing pastoral mobility in the Sahel: A question of development. Syntheses of AFC Studies and Research.

Batterbury, S. and Warren, A. (2001). "The African Sahel 25 years after the great drought: assessing progress and moving towards new agendas and approaches." *Global Environmental Change,* 11: 1-8.

Behnke, R.H., Scoones, I. and Kerven, C. (eds) (1993). Range ecology at disequilibrium: new models of natural variability and pastoral adaptation in African savannas. Overseas Development Institute, Nottingham.

Berkes, F., Colding, J. and Folke, C. (2000). "Rediscovery of traditional ecological knowledge as adaptive management." *Ecological Applications,* 10 (5): 1251-1262.

Blewett, R. (1995). "Property rights as a cause of the Tragedy of the Commons: Institutional change and the pastoral Maasai of Kenya." *Eastern Economic Journal,* 21(94): 477- 490.

Bollig, M. and Schulte, A. (1999). "Environmental change and pastoral perceptions: Degradation and indigenous knowledge in two African pastoral communities." *Human Ecology,* 27(3): 493-514.

Breman, H.A., Diallo, G., Traore, M. and Djiteye, M. (1978). "The ecology of the annual migrations of cattle in the Sahel." In D.N. Hyder (ed.). *Proceedings of the First International Rangeland Congress,* Denver, Colorado; pp. 592-595.

Brokensha, D.W., Horowitz, M.M. and Scudder, T. (1977). The anthropology of rural development in the Sahel: proposals for research. New York Institute for Development Anthropology.

Brooks, N. (2006). Climate change, drought and pastoralism in the Sahel. Discussion note for the World Initiative on Sustainable Pastoralism, IUCN. Found at: www.iucn.org/wisp-publications.

Calvosa, C., Chuluunbaatar, D., and Fara, K. (2010). Livestock and climate change. Livestock Thematic Papers. www.ifad.org/lrkm/index.html.

Coppock, D.L. (1993). "Vegetation and pastoral dynamics in the southern Ethiopian rangelands: implications for theory and management." In: *Range ecology at disequilibrium: new models of natural variability and pastoral adaptation in African savannas,* edited by R.H. Behnke, I. Scoones and C. Kerven. Nottingham: Overseas Development Institute; pp.42-61.

Coughenour, M.B. (1991). "Spatial components of plant herbivore interactions in pastoral, ranching and native ungulate ecosystems." *Journal of Range Management,* 44(6): 530-542.

Davies, J., Niamir-Fuller, M., Kerven, K., and Bauer, K. (2010). "Extensive livestock production in transition: the future of sustainable pastoralism." In: *Livestock in a changing landscape, Volume 1: Drivers, consequences and responses,* edited by H. Steinfeld, H.A. Mooney, F. Schneider, and L.E. Neville. Washington DC: Island Press.

Dyson-Hudson, N. (1966). *Karimojong politics.* Oxford: Clarendon Press.

Dyson-Hudson, R. and Dyson-Hudson, N. (1980). "Nomadic pastoralism." *Annual Review of Anthropology,* 9: 15-61.

Eckholm, E.P. (1975). "Desertification: A world problem." *Ambio,* 4(4): 137-145.

Ellis J. and Galvin, K. (1994). "Climate patterns and land-use practices in the dry zones of Africa." *BioScience,* 44(5): 340-349.

Ellis, J.E. and Swift, D. (1988). "Stability of African pastoral ecosystems: Alternate paradigms and implications for development." *Journal of Range Management,* 41(6): 450-459.

Ellis, J.E., Galvin, K., McCabe, J.T. and Swift, D.M. (1987). Pastoralism and drought in Turkana District, Kenya. Report to NORAD, Nairobi. 295 pp.

FAO (2009). *Review of evidence on drylands pastoral systems and climate change: Implications and opportunities for mitigation and adaptation.* Rome: FAO.

FAO (2006). *Livestock's long shadow: Environmental issues and options.* FAO, Rome.

Franke, R.W. and Chasin, B.H. (1980). *Seeds of famine: Ecological destruction and the development dilemma in the West African Sahel.* New York: Universe Books.

Galvin, K.A., Thornton, P.K., Boone, R. and Sunderland, J. (2003). Climate variability and impacts on East African livestock herders. *African Journal of Range Forage Science,* 21(3), 183–189.

Galvin, K.A., Coppock, D.L. and Leslie, P.W. (1994). "Diet, nutrition and the pastoral strategy." In: *African pastoralist systems: An integrated approach,* edited by E. Fratkin, K.A. Galvin and E.A. Roth. Boulder, CO: Lynne Rienner Publishers; 247 pp.

Gilles, J.L. and Jamtgaard, K. (1982). "Overgrazing in pastoral areas: The commons reconsidered." *Nomadic Peoples,* 10: 1-10.

Hidore, J.J. and Elton, M.A. (1975). "Climatic change and economic development." In: *African environment: Problems and perspectives,* edited by P. Richards. London: IAI.

IFAD (2009). Livestock Thematic Papers. Tools for project design. November, 2009. www.ifad.org/lrkm/index.html.

IPCC (2001). The regional impacts of climate change: An assessment of vulnerability. http://www.ipcc.ch.

IUCN (2011). The land we graze: A synthesis of case studies about how pastoralists' organizations defend their land rights. IUCN ESARO office, Nairobi, Kenya. 48pp.

Johnson, D.L. (1969). The nature of nomadism - A comparative study of pastoral migration in southwestern Asia and northern Africa. Research Paper No 118, Department of Geography, University of Chicago, Chicago.

Little, P.D., Smith, K., Cellarius, A.B., Coppock, D.L. and Barrett, B.C. 2001. "Avoiding disaster: Diversification and risk management among East African herders." *Development and Change,* 32: 401-433.

Lundholm, B. (1976). "Adaptations in arid ecosystems." In: *Can desert encroachment be stopped?* Edited by A. Rapp, N.H. Le Houerou and B. Lundholm. Ecological Bull. No. 24, Stockholm, Sweden.

Majok, A.A. and Schwabe, C.W. (1996). *Development among Africa's migratory pastoralists*. WestPort, CT: Bergin and Garvey.

Mapinduzi, A. L., Oba, G., Weladji, R. B. and Colman, J.E. (2003). "Use of indigenous ecological knowledge of the Maasai pastoralists for assessing rangeland biodiversity in Tanzania." *African Journal of Ecology*, 41: 329-336.

Niamir-Fuller, M. (ed.) (1999). *Managing mobility in African rangelands: the legitimization of transhumance*. Exeter: Intermediate Technology Publications.

Niamir-Fuller, M. (1998). "The resilience of pastoral herding in Sahelian Africa." In: *Linking social and ecological systems: management practices and social mechanisms for building resilience,* edited by F. Berkes, C. Folke and J. Colding. Cambridge: Cambridge University Press; pp. 250-284.

Niamir-Fuller, M. and Turner, M.D. (1999). "A review of recent literature on pastoralism and transhumance in Africa." In: *Managing mobility in African rangelands: the legitimization of transhumance,* edited by M. Niamir-Fuller. Exeter: Intermediate Technology Publications; pp.18-46.

Oba, G. and Lusigi, W.J. (1987). An overview of drought strategies and land use in African pastoral systems. Paper 23a.

OXFAM (2008). Survival of the fittest: Pastoralism and climate change in East Africa. Oxfam Briefing Paper 116.

Scoones, I. (1995). "New directions in pastoral development in Africa," In: *Living with uncertainty: new directions in pastoral development in Africa,* edited by I. Scoones. Exeter: Intermediate Technology Publications; pp.1-36.

Spencer, P. (1998). *The pastoral continuum: the marginalisation of tradition in East Africa*. Guildford and King's Lynn: Oxford University Press.

Stenning, D.J. (1957). "Transhumance, migratory drift, and migration patterns of pastoral Fulani nomadism." *Journal of the Royal Anthropological Institute of Great Britain and Ireland*, 87(1): 57-73.

Sutter, J.W. (1987). "Cattle and inequality: herd size differences and pastoral production among the Fulani of north-eastern Senegal." *Africa*, 57: 196-218.

Swallow, B. (1994). The role of mobility within the risk management strategies of pastoralists and agro-pastoralists. IIED, Gatekeeper Series, No 47.

Swift, J. (2000). "Prospects for the Sahelian pastoral economy." In: *The Sahel: energy supply, economic pillars of rural Sahelian communities, and need for revised development strategies.* Proceedings of the 12th Danish Sahel Workshop, 3-5 January 2000. SEREIN Occasional Paper, No 11; pp. 77-88.

Warren, A. (1995). "Changing understandings of African pastoralism and the nature of environmental paradigms." *Transactions of the Institute of British Geographers,* NS 20: 193-203.

Westoby, M., Walker, B. and Noy-Meir, I. (1989). "Opportunistic management for rangelands not at equilibrium." *Journal of Range Management,* 42: 266-274.

World Initiative for Sustainable Pastoralism (WISP) (2007). "Change of wind or wind of change?" Climate Change, Adaptation and Pastoralism.

CHAPTER 8

Nomadic Pastoralism and Natural Resource Use Conflicts in
East Africa

Paul Onyango

Zoology Department, University of Dar es Salaam, Tanzania

Introduction

"Pastoralism is a highly flexible system... (which) ...has evolved
over time as the most efficient means of exploiting short-lived water
under naturally marginal conditions, and prevailing technological and
economic situations" (Summary of Umar, 1994, as quoted from Omosa
2003). Nomadic pastoralism is an activity in which those involved
primarily depend on animal husbandry in arid and semi-arid lands.
Pastoralists have to find adequate grazing land and water for their
animals, thus they move from one place to another frequently (Hunter,
n.d.). This kind of economic activity has remained a lifestyle for a
number of communities throughout the world for a long time.

Pastoralists occupy about a quarter of the world (Desta, 2006). This
includes the dry lands of Africa (where 66% of the total land area is
under pastoralism) and the Arabian Peninsula, to the highlands of
Asia and Latin America. They raise domestic animals, including cattle,
camels, goats, sheep, and donkeys, which are used for milk, meat,
transport, and trade. Africa alone accommodates over 60% of the world's
pastoral peoples where thirteen million are predominantly pastoral,
while another nine million are agro-pastoral, keeping large numbers of
livestock as they practice agriculture (Desta, 2006; Galaty and Johnson,
1990; Jahnke, 1982).

In East Africa, pastoralists occupy more than half of the land area of the three East African countries. For instance, they "occupy 70% of the total land of Kenya, 50% of Tanzania and 40% of Uganda" (Fratkin, 2001). They are, however, numerically few (*i.e.* less than 1.5 million in each of the East African countries) (*ibid.*).

Pastoralists occupy savanna, semi-arid, or arid lands where rain-fed agriculture is quite a difficulty. In Africa, pastoralist communities include the Tuareg, the Fulani (also known as Fulbe or Peul), the Arrial, the Bedouin, the Somali, the Nuer, and the Maasai, among others (Fratkin and College, 2004; Adriansen, 1997). Typically, these communities occupy large tracts of communally shared land and use kinship ties for herding and defence. Their herds are often large and adapted to their environment in which they have to survive periodic droughts and sparse vegetation.

Several pastoralist communities practice some agriculture, particularly planting of maize, millet, or in Ethiopia what they call *teff* (Fratkin, 2001). They also supplement their pastoral diets with wild plants, game, and fish obtained from long distance trade (Fratkin, 2001).

They are politically and economically disadvantaged compared to other ethnic communities in the countries where they are located. For example, the areas they live in are prone to unpredictable rain and periodic droughts. They are vulnerable to food insecurity and health problems and in some cases they have low command of the national language of their countries thus reducing their ability to access health services as well as education (Lawson *et al.*, 2014).

They are in constant migration in search for pastures. Such a migration routine has had an effect on their advancement, especially with regards to education, technology and even business development. Moreover, development models that focus on migrating communities are yet to be formulated. As pastoralist communities migrate, they meet sedentary communities. For a long time, such movements have resulted in overuse of the natural resources, such as, for instance, clearing of areas which were once good grazing lands and reduced availability of water.

While natural environments have been cleared, climate change has also affected these natural resources. This effect has led pastoral communities to move to areas which are traditionally outside their migration routes. They have moved to villages where other farming communities have settled down. Their movement to these farming areas has led to conflicts as they seek to use land resources which are the life line for the resident farming communities.

Climatic change has become a factor in conflict considerations when unanticipated environmental changes emerge to alter existing land tenure arrangements between and among communities. More often than not, these changes have resulted in pastoral communities encroaching on the land of their neighbours, a tendency that has been violently resisted by the aggrieved farming communities. Also important in this consideration is when droughts have affected grazing land which pastoralists use for their animals, and such animals have trespassed on agricultural farms (Alao, 2011). It is the exploitation/use of natural resources that leads to these conflicts.

Umar (1994) argues that conflicts emerge as one way in which human society adjusts in the face of scarce resources. Conflicts are complicated issues that get interwoven with other social, economic, environmental and political activities within a society. If not managed well, conflicts can have a negative impact on the environment and on the assets and capabilities people require for their survival.

As a consequence of the rising conflicts between several groups worldwide, there has been considerable interest in them, both because civil conflicts retard development and because their causes and origins are largely unanticipated. It is also the case that exaggerated and highly-sensitized media accounts tend to escalate misunderstandings between and/or among groups (FAO, 2000). In addition, understandings about conflicts get guided by such media perceptions to an extent that a sober approach to conflict resolution is lost. Thus against this background, it is important to generate literature that will focus on the understanding of natural resource conflicts involving pastoralists against other groups. As a consequence, there have been a number of missions, studies and workshops to address the issue. Not all of these have been productive; and indeed by simply reprising old literature of preconceived ideas, they are unhelpful.

This chapter, therefore, discusses these conflicts. In particular, it examines the types and nature of conflicts, areas where conflicts have been highly reported, sources of and main reasons for the conflicts, key features of pastoralism, and natural resource related conflicts and their root causes. The chapter also introduces an interactive governance framework useful in understanding conflicts among pastoralists and how they can be assessed and addressed.

Understanding Natural Resource Use Conflicts

A conceptual framework

Natural resource use conflicts come in different ways in different places. In some areas, conflicts arise due to differences over access to and use of natural resources by two or more individuals and/or groups. These conflicts often emerge because individuals and/or groups have different uses for resources such as forests, water, pastures and land, or because individuals want to manage them in different ways. Disagreements also arise when these interests and needs are incompatible, or when the priorities of some user groups or individuals are not considered in policy formulation and implementation of programmes and projects. Such conflicts are an inevitable feature of all societies.

In recent years the scope and magnitude of natural resource use conflicts have increased and intensified. Thus, if not fully addressed, they can escalate into violence, causing environmental degradation, disruptions of projects and undermining of livelihoods. Acknowledging that conflict is a common feature of any resource use system is also a prerequisite for sustainable natural resource management.

There are basically two schools of thought providing a theoretical understanding of natural resource use conflicts. On the one hand, one school of thought argues that conflicts occur due to scarcity of natural resources. As in the case for historical China, Pei and Zhang (2014) clearly show that resource scarcity, which "…includes reductions in the quantity or quality of resources" due to the impacts of climate change that degrades grasslands, "…will definitely increase the difficulty of making a living for pastoral societies". This scarcity "…will not only shrink the whole pie for society, it will also decrease the slices for each individual as it increases their vulnerabilities."

This school of thought is dominated by the neo-Malthusian scholars who claim that rapid population growth, environmental degradation, resource depletion, and unequal resource access combine to aggravate poverty and income inequality in many of the world's least developed countries (Kahl, 2006). These deprivations are easily translated into grievances, increasing the risks of rebellion and societal conflict.

The other school of thought argues that conflicts occur due to natural resources being overabundant. This school argues that resource abundance, rather than scarcity, provides the basis for understanding natural resource conflict. Some countries with abundant natural resources have experienced what has been coined as the "resource curse", with corruption and economic stagnation resulting in violent conflict over access to revenues from resources. The availability of easily looted natural resources, without proper and clear demarcation and ownership has encouraged conflicts to grow in many African countries (USIP, 2007).

Whether natural resources are in abundance or in short supply, this chapter takes the view that conflicts arise and can be understood through a governance lens or, in particular, through an interactive governance lens. The 'access' challenge when natural resources are in short supply and the 'resource curse' when they are in plenty all indicate that the whole issue has to do with governance. In other words, governance should not be perceived here in a technical sense involving tools that can be applied to address a tangible assignment where the objective is clear and the result measurable. Governance is argued here to be broader than management. It includes deliberations among those being governed and those governing in formulating goals to be achieved (Jentoft and Chuenpagdee, 2009). It is on the need for participatory decision making that the concept of interactive governance is grounded.

The interactive governance framework provides a means of understanding the extent to which the existing form of interaction gives actors in the conflict opportunities to influence relevant policies to resolve their conflicts through mobilizing their practical knowledge and social values. The framework also provides a means of understanding how conflicts arise and are sustained. Indeed an explanation for the persistent and unending conflict can be seen as evidence that conflict resolution strategies are not congruent with local level institutions and thereby not meeting their intended goals.

According to some theorists (Chuenpagdee and Jentoft, 2009; Jentoft and Chuenpagdee, 2009; and Bavinck et al, 2005), the framework operates in terms of three aspects, namely:

a) systems-to-be-governed (which includes a natural and a social system),
b) a governing system, and
c) a governing interaction.

The systems-to-be-governed comprises the communities involved in the conflict as well as the natural resources in dispute. The governing system comprises the government, civil society institutions (including development partners, private sector and or market) and the community organizations involved in governing. The interaction between these two systems is central in understanding how conflict comes about and how effective resolution strategies are and can possibly be governable (Kooiman and Jentoft, 2009; Kooiman, 2003). The two systems and their interactions are within this framework seen as inherently diverse, dynamic and complex and operate at different scales (Chuenpagdee and Jentoft, 2009; Jentoft and Chuenpagdee, 2009; Bavinck *et al*, 2005; Kooiman *et al*, 2005; and Kooiman, 2003). These systems' characteristics create limits and opportunities for conflict resolution.

To understand conflicts using this framework, it is argued that conflicts occur, and even persist, as a consequence of ineffective interaction between the two systems. This is the case, for instance, where actors have unequal opportunity in influencing outcomes. The interactions mainly revolve around decision-making processes, power relations, representation, popular participation and learning. The inadequacy and ineffectiveness in interaction between the two systems is an issue that can partly be traced from the values that the actors within the two systems hold. This is because values determine how governors perceive conflict and how they prefer to deal with it and how the communities in conflict respond.

In other words, the sociocultural context of values, norms and principles is important in understanding community actions that perpetuate conflict as well as try to resolve it. Members of a community are always in a dynamic psychological field. They interact within and without their communities and through this they create a social plane. It is on this social plane that conflict occurs. As Rummel (1976) pointed out, society and culture infuse the dynamic field through:

1. a cultural matrix which gives stimuli perceptual interpretation,
2. the social roles which are part of attitudinal lattice,
3. the bounding and definition of behavioural potentials, and
4. a structure of expectations.

Thus cognition and perception of reality are organized in terms of cultural meanings, values, and norms. Our world is intrinsically subjective; physical objects serve only as vehicles of meanings and values.

Conceptually, cultures are a system of meanings, values, and norms which vary along language, philosophy-religion, ethics-law, science, and fine arts components (Kooiman and Jentoft, 2009; Kooiman and Chuenpagdee, 2005; Rummel, 1991). These components span the space of cultural potentialities in which the world's cultures cluster into materialistic and/or otherworldly types, depending on their particular system of meanings, values, and norms. Moreover, societies share the system of meanings, values, and norms of their culture and in addition add the status components of wealth, power, prestige, and class.

In total, the whole that is the intentional space consists of a bio-psychological, social, and cultural space, and the spaces of the environment and vehicles. The intentional space in humans can be seen in the componential structure, which defines motivations, temperaments, abilities, moods, and states. In addition, as Rummel (1976) emphasizes, there are the meanings, values, and norms components of religion-philosophy, science, language, ethics-law, and fine arts; the status components of wealth, power, and prestige and the component of class, and finally the four space-time components of the physical world.

At the centre is the dynamic nature of human beings, motives, attitudes, sentiments, and above all, superordinate striving for self-esteem and self-actualization. Values, therefore, influence attitudes and behaviour because they determine norms, principle and choices for any action to be taken. This implies that values, norms and principles are inherent in social action and they differ from one culture to another. Hence, when a community lists factors for which they have to fight, these reasons are value-laden and are often specific to the community. The reasons will reflect the understanding of conflict in the communities involved.

Systems–to-be-governed: Areas and reasons for pastoral conflicts

The natural system

Within the interactive governance framework, the natural system of the systems-to-be-governed comprises the natural resources in their diverse, dynamic and complex state. These characteristics are often overlooked when solving societal problems (Kooiman *et al.*, 2008). With respect to conflicts among the pastoralist communities, the natural system involves land, water, grazing areas, arid and semi-arid areas. Unfavourable climatic conditions also play a critical role and often precipitate the need for pastoralists to migrate in search of pasturelands and water. During the dry season, sources of water become desiccated and lands turn arid. This forces young pastoralists to leave villages with their livestock in search of water and grazing lands. It is during these times that pastoralists engage in conflict over herding territory and replenishment of lost cattle.

When conditions are made worse by extended periods of drought, conflicts become more intensified and frequent (Oxfam, 2006). In the past few decades, drought-related emergencies have risen sharply. For instance, between 1975 and 2006 the number of people affected by drought rose from 16,000 to an estimated three million (*ibid.*). This calculation does not take into account that the population has also grown significantly, but not nearly as exponentially as those impacted by the recurrent droughts. These changes in drought signify a dynamic natural system that governors have to confront.

Availability of water during dry seasons has been a cause of conflicts between pastoralists and farmers. The presence of water has meant that agriculture can do well, but at the same time it has meant that grass is available for pastoralists' herds. Conflicts have, therefore, been observed between pastoralists and farmers in areas where water is found, especially during the dry periods. The bone of contention here has been land use. Pastoralists argue for grazing areas while farmers have argued that the former have taken up their rich and fertile farmlands. Conflicts emanate partly from the increasing demand on these resources because of increase in population (both human and bovine), continued depletion of these resources in both quantity and quality due to degradation, overuse and over-harvesting, ineffective governance and external factors such as climate change and commercial pressure.

In some areas conflicts arise when neighbouring communities claim land in their shared border. A good example of such a conflict would be that between the Garre and the neighbouring Borana in Ethiopia (Odhiambo, 2012). These communities have historically contested ownership of the land that separates them with each claiming that the area belongs to them.

In the recent past, East African pastoralists have faced large challenges in their economies and traditional ways of life. For instance, in the savanna regions of southern Kenya and Tanzania, the Maasai and other pastoral communities have lost land to private ranches, wheat estates, and the expansion of protected areas. National governments, with very good intentions to increase investments, have increasingly privatized formerly communal grazing lands, encouraging the expansion of exports of agricultural and livestock products as well as industrialization.

In the more arid regions of northern Kenya, northern Uganda, and southern Ethiopia, pastoral populations such as the Samburu, Pokot, Turkana, Rendille, Boran, and Karimojong have faced problems of drought and famine coupled with ethnic conflict and political insecurities, exacerbated by population growth and increased competition for range land and water resources. Inter-ethnic fighting has increased violence as weapons from civil wars in Sudan, Ethiopia, and Somalia are increasingly acquired by pastoral tribesmen (Plate 7.1) (Fratkin and Wu, 1997; Little, 1992; Spencer, 1998).

Photograph 8.1: *A young Borana pastoralist wields a Kalashnikov gun. Weapons from civil wars in Sudan, Ethiopia, and Somalia are increasingly being acquired and used by pastoral communities in pastoral conflicts across the East African Region.* (Source: Hesse and Cavanna, 2010).

In Tanzania, land conflicts have mainly involved pastoralists and conservationists, crop cultivators and pastoralists, government and villagers, villagers and investors. These have mostly occurred in Morogoro (particularly Kilosa District), Mbeya, Arusha and Mara regions (Benjaminsen *et al.,* 2009; Olengurumwa, 2009). These land-use conflicts have caused harm to community lives and animals, destruction of properties, insecurity, low productivity, accelerated food insecurity, generation of a landless class, increased poverty, and rapid environmental degradation (NGONET, 2008; Verstegen, 2001).

In Ngorongoro District, Tanzania, the Maasai have protested the expansion of the Ngorongoro Conservation Area (NCA) and adjacent protected areas. Such expansion has reduced available land for the Maasai to graze their animals. The NCA, on the other hand, have demanded more land for environmental conservation and wildlife protection in a World Heritage Site (UNESCO, 2007). The Maasai community have been forced to leave their local areas for areas where there is not enough water and pastures for their livestock (Michael, 2008). This is exacerbated by lack of land-use plans at the community level and within the NCA; and even if the plans are in place, they are normally not implemented, thus leading to ongoing conflicts over land use for agriculture and livestock production.

The social system

Changes in natural resource characteristics have also brought about changes in the livelihoods of the pastoral communities. These changes can be traced to how these communities have interacted with the natural system. Dependence on the natural system for grazing and demand for water has had a great impact on the lives of these pastoral communities. In response the communities have designed new strategies for survival in the wake of these changes. In applying these new strategies, conflicts have been encountered, largely caused by competition over control of and access to natural resources, particularly water and pasture. Other causes of conflicts have included historical rivalry, deep-seated cultural differences, land tenure issues, political incitements, and more recently idleness amongst the youth and proliferation of illicit arms (USAID, 2005).

Conflicts in Kenya, for instance, present themselves in many different forms. Juma (2000) classifies them into four main categories:

a) conflicts among pastoral communities themselves,

b) conflicts linked to presence of refugees,

c) conflicts between pastoralists and crop farming communities, and

d) ethnic clashes.

The author observes that pressure on the environment has more often than not changed conflict dynamics. For example, the 1999-2001 droughts that wiped out the Turkana community's entire stock increased pressure on the Turkana to raid other communities' herds in order to restock their lost herd. The drought of the year 2011 in the arid northern Kenya depleted pasture and dried water points in Moyale, triggering inter-clan attacks and counter-attacks (IOM, 2011).

Impacts of drought have equally led to conflicts in Tanzania. Provision of hunting blocks to companies is another area which has led to perennial conflicts. For example, companies permitted by the conservation authorities to hunt different animals in Loliondo area have placed restrictions on the use of grazing estate of the Maasai and thereby reduced the latter's mobility. This has led to a serious conflict between the local Maasai community and the conservation authorities (Shivji, 1993).

Conflict is also not a new phenomenon in arid north-western Kenya. For centuries pastoralist communities have periodically used violence to access land, water and to raid livestock from neighbouring groups (Eaton, 2008). However, in recent decades the cultural practice of livestock theft has experienced significant changes. Combining the availability of small arms with the effects of extended droughts, raiding has become more frequent, more violent and more destructive (GoK, 2008; Mkutu, 2008; UNDP, 2011; UNOCHA, 2011; Schilling et al., 2012a; Schilling et al., 2012b).

Almost 60% of all the raids in Turkana County in Kenya between 2006 and 2009 were conducted by the Pokot of Kenya and Uganda, followed by the Toposa of Sudan (11%) and the Dasenach of Ethiopia (9%) (TUPADO, 2011). In the southern part of Turkana, the Tepeth and Jie of Uganda also raid the Turkana on their own or in alliance with the Pokot (Lokuwam, 2011; Mukoo, 2011; Raider, 2011a). The Turkana, on the other side, raid the Pokot in West Pokot and less often in Uganda. Occasionally, the Turkana form raiding alliances with the Matheniko of Uganda to attack the Pokot (Mukoo, 2011).

The governing system

The governing system involves the rules, institutions and organizations involved in addressing conflicts among the pastoralists. These include both the government, and non-state actors such as the private sector and the civil society. The governing system is also diverse, complex and dynamic policy wise, institutional wise and other socio-economic wise, as discussed below.

1) Colonial and post-colonial policies: Deficiencies and contradictions

Cattle raiding among pastoralists is a phenomenon that stretches back many centuries. Traditionally, cattle rustling, often involving some violence, was redistributive and only involved the theft of cattle to replenish herds after decimation from drought or need to pay out bride price. When ethnic group members were killed, cattle were offered as compensation and the culprits were subjected to intense cleansing rituals. Prior to the system of hierarchical government, councils of elders, traditional courts, and peer groups were at the centre of authority among tribes. As such, they governed raids to ensure that they did not spiral out of control, and when disputes arose, traditional mechanisms were employed to settle them (Mkutu, 2003). Typically, there was some loss of life from raiding, but on a much smaller scale.

Colonial rulers disrupted the pastoralist social order, replacing it with a system of provincial government appointees within newly established borders that limited the free movement of pastoralists with their animals. Traditionally, land belonging to families was passed down from one generation to the next, but alienation of pastoralist land, combined with discriminatory land reforms eroded this custom (Kandagor, 2005). In addition, pastoral communities were isolated from other areas that enjoyed the benefits of colonial security and development (Mburu, 1999). The apparent crackdown on cattle raiding is emblematic of an overall attack on pastoralism itself, on the grounds that it was a primitive and thus inhumane way of life. The weakening of traditional governance systems has thus undermined pastoralist authority and ability to settle disputes. Without adequate alternatives to replace traditional structures of governance and security, pastoralists operate in an anarchic environment (Mkutu, 2003).

In Tanzania, the Land Policy, through the Land Act and Village Land Act of 1999, classifies land as reserved land, village land, and general land. Reserved land is statutory, protected as national parks, land for public utilities, wildlife and game reserves and other land designated by sectoral legislation. Village land is land which is within the demarcated or agreed boundaries of any of Tanzania's recognized villages. This land is under the authority of the Village Government, which is answerable to the Village Assembly.

General land is a residual category and includes all public land which is not reserved land or village land and includes unoccupied or unused village land. The definition of General Land is ambiguous because unoccupied or unused village land is considered as "excess" and thus falls under the jurisdiction of the Land Commissioner rather than the village authorities. The seemingly unoccupied lands (village) may be important areas for seasonal livestock grazing, and other important livelihood uses (Mattee and Shem, 2006). Certainly this ignores the fact that as the population grows this "excess" village land will eventually be brought into use. However, this is actually the same land which is, in most cases identified as "Land Bank" suitable for agricultural investment.

Not coincidentally and under the pretext of large scale agricultural investments, the land belonging to farmers and pastoralists is subject to appropriation by state-backed investors. Thus, the pastoralists occupying semi-arid areas are often subjected to efforts to alienate their customary pastures and land holdings, for purposes of commercial investments or establishment of wildlife conservation areas (Mattee and Shem, 2006). The land policy has some deficiencies because it does not guarantee security of tenure to some users, especially to pastoral groups. In effect, these deficiencies have led large areas of land being given over to alternative uses and consequently marginalizing the pastoral populations (Bonfiglioli, 1992; UDSM, 2015).

Appropriation of land from pastoralists in Tanzania, like elsewhere in Africa, is usually backed up by an enduring perception that pastoralism is an irrational, ecologically destructive and economically inefficient production system (Homewood, 1995; Hesse and MacGregor, 2006). These perceptions have consequently resulted in efforts by government policy makers to re-distribute pastoral lands directly to commercial investors in the belief that this is an economically rational policy (Sulle and Nelson, 2009). The squeezing out of herders to give way to large scale seed farms in Arusha and Manyara regions in northern Tanzania typifies the argument here and bears witness to the limits of policy makers' understanding of the nature of pastoralism. Suffice to point out that as long as the pastoralists continue to be squeezed out of their traditional grazing lands, migration into other parts of the country in search of pastures and the subsequent conflicts cannot be avoided (ibid.).

In addition to policy deficiencies, there is also a problem of contradicting policies. This is very well supported by Lugoe (2011), who clearly shows, for example, that there is some misalignment between the Livestock Policy of 2006 and the National Land Policy. The Livestock Policy recognizes seasonal movement as an important characteristic of pastoralism and thus encourages livestock owners in overgrazed areas to move to lower stocked areas. The Livestock Policy has gone further and facilitated modalities for new settlements for pastoralists. In contrast to this spirit, the National Land Policy prohibits nomadism and all its different forms - modern or transhumant. Such contradictions help to sow seeds of hostilities between the pastoralists and implementers of the policies.

2) Inadequate capacity of the local institutions

The influx of livestock into areas which were once dominated by crop cultivators has contributed to the occurrence and persistence of conflicts between farmers and herders. This has been an inevitable outcome because the cultural values and attitudes which used to provide the basis for interaction and the norms by which individuals and communities live have been altered. At village level, the traditional conflict resolution machinery has been weakened partly by the emergence of statutory approaches based on formal procedures, and on the other, by the influx of herders who do not share the values and beliefs upon which these mechanisms were anchored.

Elsewhere in Sub-Saharan Africa, it has been also noted that land conflicts are proving more difficult to resolve because traditional instruments of conciliation such as compromise and consensus are failing. On the one hand, while local institutions that used to arbitrate such conflicts have largely lost their authority, few institutional innovations have been developed to play the roles of arbitration (Kirk, 1999).

The good news is that at the national level, various local and international non-governmental organizations (NGOs) are becoming involved in peace building. Such peace building initiatives include: the Development Initiative Access Link (DIAL) Africa in Somalia (DIAL, 2012), the Greater Horn of Africa Initiative (GHAI) for the Greater Horn of Africa (GHA) (USAID/REDSO, 2003), the Canadian International Development Agency (CIDA) and the UNDP Reduction of Resource Based Conflicts Project which was launched in 2004 in Sudan (UNDP Sudan, 2010).

3) Loss of common property resources

Whereas livestock among most East African pastoralists constitutes individual or family property, access to land (for pasture, water, minerals, and security) is usually shared by territorial or kinship groups (*i.e.* land is held in common as a communal resource) or it is considered common property open to all. Ignoring traditional land tenure in favour of individual tenure rights, Kenya, Tanzania, and Uganda have encouraged the privatization of communal lands, following policies encouraged by international donor organizations including the World Bank and USAID. For example, following the establishment of 'ranching associations' in Tanzania and 'group ranches' among Maasai in Kenya in the 1960s and 1970s, the Kenyan government is now promoting private and individual titles, leading to a scramble for land similar to the one that took place in the U.S. West in the nineteenth century (Galaty, 1994). In the 1980s the Tanzanian government, with assistance from the Canadian International Development Agency (CIDA), developed large parastatal wheat estates around Mount Hanang (NAFCO), displacing Barabaig herders from 100,000 hectares of their land.

In addition to land loss to agriculture, pastoralists have also lost large areas of land to national parks such as Amboseli, Maasai Mara, Tsavo, and Samburu Parks in Kenya, and the Serengeti, Ngorongoro Crater, and Mkomazi in Tanzania (Homewood, 1995; Brockington, 1999).

4) Commoditization, sedentarization and urban migration

Pastoralists have increasingly shifted their economy from subsistence production (producing mainly milk for household consumption) to commercial production (producing beef and dairy products for sale both to domestic and export markets). The sale of livestock is not new to pastoralists. The Tanzanian Maasai were trading livestock for grain with neighbouring Arusha in the mid-nineteenth century (Spear, 1997). It is estimated, for example, that the Maasai have typically sold between 8 to 10% of their cattle to purchase grains and other commodities since the 1930s (Zaal and Dietz, 1999).

Increased commoditization of the livestock economy has led to a growing polarization of pastoralists into haves and have-nots, particularly in Maasai areas that are close to urban markets. This economic differentiation has also occurred among the Boran, the Rendille, and other pastoralists becoming integrated into the market economy. While those households with large livestock herds have remained committed to the pastoral economy, poor pastoralists have worked for wealthier kinsmen, adopted agriculture, or migrated to towns in search of low-paying jobs, working as watchmen, maids or prostitutes (Talle, 1988; Mung'ong'o, in this book).

Governing interactions: Conflict mitigation approaches

1) Conflict mitigation approaches

This section analyses and evaluates three conflict mitigation approaches: peace meetings, disarmament, and facilitation of secure livestock movement. Lind (2006) defines peace building, as "...any activity that is undertaken with the purpose of preventing, alleviating, or resolving conflict, ... is a regular practice amongst the warring communities". Furthermore, Lind (2006) observes that much of the donor agencies' activities in pastoral areas of eastern Ethiopia, northern Kenya, south-eastern Sudan and northern Uganda revolve around "conflict reduction and peace building". At local levels, disputes over lost crops and access to water and pastoralist routes are resolved by councils of elders (UNDP Sudan, 2010).

Between October 2009 and September 2011, several peace meetings were organized by international and local NGOs such as the Practical Action, Reconcile, and International Rescue Committee under the European Instrument for Democracy & Human Rights (EIDHR) (Locham, 2011). The biggest meeting was the 38th Peace Accord held in Lokiriama on 21st September 2011. The meeting was attended by community members from the Turkana and Pokot of Kenya, the Matheniko and Jiie of Uganda, the Topossa of Sudan and the Nyangatom of Ethiopia. The aim of the Accord was to "promote peace, human security and development in the region through the use of local peace initiatives by the communities in Karamoja cluster" (Practical Action, 2011).

Meetings between conflicting parties generally offer the opportunity to mitigate conflict through interaction and dialogue. At the Lokiriama Accord, however, no discussion was facilitated among the government representatives or participants. A second critical question is: who is attending the peace meeting? Smaller peace meetings are usually attended by local leaders and members of the peace committees which consist of elders or other respected members of each community (Archibo, 2011; Lodea, 2011; Sikirika, 2011).

One elder who is regularly called to the peace meetings summarized the success of these meetings in a few words "They don't help." (Sikirika, 2011). This poor result could be probably due to the fact that such processes are not home grown, that they are externally driven (Lodea, 2011; Archibo, 2011; Raider, 2011b; and Sikirika, 2011). On the other hand, not all stakeholders are represented. Women, for example, play a very peripheral role in the peace meetings. As one interviewee phrased it: "So far, the few women attending the meetings only prepare tea for the men" (Tulel, 2011).

Against this background, it seems recommendable to let elders and women facilitate the dialogue between community members and especially the raiders. To minimize the raiders' fear of punishment, the attendance of governmental representatives could be limited. This approach is promising in building long lasting peace agreements which have the potential to stabilize relationships between groups, as the Peace Agreement between the Turkana and the Matheniko has shown for almost 40 years (Lomor, 2011; Sikirika, 2011).

2) Disarmament campaigns

Efforts at disarmament are crucial in mitigating conflicts. There can be little doubt that the availability of small firearms has made conflicts in north-western Kenya more deadly (Mkutu, 2006). The Ugandan and the Kenyan governments have made numerous non-forceful and forceful attempts to disarm the region (for example in 2001 and 2009; and in 1984 and 2002, respectively) (Knighton, 2003; Mieth, 2006; Eaton, 2008; Kaliamoi, 2011; Longolol, 2011; US Department of State, 2011). The disarmament campaigns of the 20th Century were unsuccessful in the best case and they led to massacres in the worst case (Wepundi *et al.*, 2011). The government of Kenya established the Kenyan National Focal Point on Small Arms and Light Weapons in 2003 to improve the coordination of disarmament operations (GoK, 2011). The most recent disarmament campaign *Dumisha Amani II* ("Sustain Peace") in north-western Kenya started with a voluntary phase in February 2010, which was followed by a coercive phase two months later (Raider, 2011a&b; Wepundi *et al.*, 2011). Safer World, citing government records, states that until August 2010, the campaign had recovered 1,201 firearms (*ibid.*).

Disarmament is a highly delicate instrument. If done selectively it can make "one community more vulnerable to others who are still armed" and therefore "lead to extinction of some communities" (Interview by Moses Akuno, as quoted in UNOCHA, 2010a:4). It has to be done in all communities equally and taking into consideration overarching concerns and standards. The exercise should be seen as legitimate according to some moral principles and values. For disarmament to work, it has to be done in all relevant communities simultaneously and to the exact same degree.

3) Facilitation of secure livestock movement

Mobility is an essential part of pastoralism (Birch and Grahn, 2007). To enable the pastoralists to adapt to climate change, a group of international agencies formulated what has been called a Security in Mobility (SIM) initiative. This was started "to promote pastoralist mobility as a climate change adaptation strategy, to advocate for the reconciliation of regional security concerns with pastoralist livelihood and mobility needs, and finally, to support governments in the region to develop a regional normative framework on safe migration for pastoralists" (UNOCHA, 2010b:13). The SIM initiative addresses this issue of mobility which is relevant and widely neglected. The different backgrounds and expertise of the participating agencies ensure that the problem is analysed from a variety of perspectives. The SIM initiative also addresses political marginalization, resource scarcity and conflict as key challenges.

In response to resource based conflicts in Tanzania, various stakeholders, including the government, international organizations, local NGOs and CBOs have made several efforts to combat the problem. These have included preparation of land-use plans and establishment of policies on pastoralist natural resource management, as well as introduction of Community Based Conservations (CBC) to ensure that pastoralists benefit from conservation and that the communities are directly involved in conservation issues (Wright, 1993; Western *et al.* 1994; Adams and Hulme, 2001; Hackel, 1999; Hulme and Murphree, 2001; Mattee, 2007). Despite all of these initiatives made by the government and other stakeholders in Tanzania, land-use conflicts still persist in areas such as Ngorongoro Conservation Area and surrounding protected areas, Kilosa and other districts.

4) Pastoralists' political marginalization

Both prior to and after independence, pastoralists have been side-lined in decision-making processes in most East African countries. The result is chronic socio-economic underinvestment in pastoralist communities across the region. The cause of this marginalization is in part geographical. Pastoralist communities tend to be remote and highly mobile. Ethnicity is also another factor that has fuelled marginalization. Most of the major pastoralist groups move across national boundaries: for instance, groups in the areas between Ethiopia, Eritrea, and Djibouti; the Somalis between Somalia and Kenya; the Borana between Ethiopia and Kenya; and the 'Karamoja cluster' between Kenya, Uganda and Sudan. These cross-border identities render pastoralists vulnerable to the political cultures of the various nation states (Morton *et al.,* 2007).

Pastoralists are sometimes believed by their fellow nationals to have divided loyalties, and are highly vulnerable when such accusations suit other political interests. However, perhaps the greatest source of pastoralist marginalization is the outdated idea, which dominated much of the development thinking in the latter part of the twentieth century and in many areas continues today: that pastoralism is an outmoded way of life that needs replacing with 'modern' livelihood systems (*ibid.*).

Governments in the region have historically had little economic and political interest in promoting pastoralists' interests, as they have tended to see pastoralists as a 'minority vote' that is not worth winning. In several East African countries, pastoralists are relatively few in number and occupy what is considered by their governments to be marginal lands with little economic potential. Even in countries where pastoralists are the majority such as Somalia and Somaliland, political power is concentrated in the hands of an elite which tends to use it to pursue their own short-term political and economic agenda, rather than for the common good of the majority, which includes pastoralists (Oxfam, 2008).

Despite having some strong traditional institutions, in general the pastoralist areas are politically weak and disorganized, due to their social and economic marginalization and governments' rejection or misunderstanding of their traditional systems of authority and leadership (ibid.). The creation of conservation areas has led to pastoralist land loss (Behnke, Scoones and Kerven, 1993). In Tanzania, for example, conservation areas have led to more land being taken from pastoralists. 95% of Monduli District, which is at the heart of Maasailand, has been set aside for conservation. Countrywide it is estimated that more than one-third of protected areas have traditionally belonged to pastoralist communities (Olenasha, 2004). Over the past half-century, Maasai pastoralists in Tanzania have been evicted from much of their lands in the name of conservation. Large-scale evictions occurred in the Serengeti in 1958/59, in Mkomazi in 1988, and in Ihefu (Mbarali) in 2006 (ibid.). In Uganda there is a similar situation. 70% of the land in Karamoja has been allocated as hunting grounds or protected areas such as Kidepo National Park. Generally, pastoralist communities are not secure in the land they inhabit (Karani, 2008).

Features of Natural Resource Related Conflicts

Propensity to maximize herds for cultural and economic reasons

Nomadic livelihoods are based on livestock and their products, which can be consumed or sold. Nomadic pastoralists often cover great distances with their livestock, following pasture availability (Blench, 2001). The majority of nomadic pastoralists does not have permanent settlements and consequently use mobile homes such as tents (Blench, 2001). African pastoral societies are characterized by a cultural and economic orientation towards livestock, be it cattle, sheep, goats or camels. Any surpluses generated by the pastoral system tend to be reinvested in livestock with a view to herd maximization.

Larger herds guarantee subsistence and income, confer status and may provide insurance against the impact of drought. Even educated members of pastoral ethnic groups, no longer dependent upon pastoral production for their livelihoods, tend often to continue to invest in livestock, often as absentee owners. In good times there is a surplus of animals. The extent to which these are marketed depends on regional and inter-ethnic variation, interacting with the availability of and access to markets (Kerven, 1992). A key feature of African pastoralism is mobility, with herds being able to move to distant grazing along traditional migration routes or in time of emergency to areas where grazing or browse is available (*ibid.*).

The marginal nature of pastoral environments and vulnerability to drought mean that mobility of people and livestock is essential. Herd growth is, therefore, opportunistic and livestock numbers only increase during good years. Occasional crashes in the livestock population are inevitable in the face of drought and herd reconstitution post-drought can be a slow process depending upon the numbers of breeding stock which survive a drought. During droughts, pastoralists are particularly vulnerable to fluctuations in terms of trade between livestock and grain products as livestock prices plummet and grain prices increase (Toulmin, 1995).

Pastoralists keep animals for several reasons. They are a form of productive capital, they provide for subsistence in the form of milk, meat and blood, provide transport in the case of oxen, donkeys and camels, and serve as an important store of wealth and insurance. As a form of insurance, they may be imperfect as they are subject to the impact of drought and disease. However, in the absence of alternatives, particularly financial markets and institutions, they are the only form of insurance available to many pastoral households (Hogg, 1997).

In some cases, the pastoralist capital is consumed directly as slaughter, but more often animals are sold to realize cash for grain purchases and other necessities of life. The store of wealth in animals is also reflected in social institutions such as marriage and inheritance. The largest transfer of livestock a male is likely to make in his lifetime is for bride wealth at marriage. Livestock are, therefore, a source of prestige and a means for partaking in complex networks of social obligations and reciprocity that mitigate risk, particularly for poor households (Hogg, 1997).

According to Benjaminsen *et al.* (2009), "(p)astoralists have, since the colonial times, been perceived by authorities and development organizations in Africa as unproductive .... unorganized .... and environmentally destructive (they cause overgrazing and desertification) and contributing nothing to national economies". They have also been conceptualized as unorganized ('roam around') bands, and environmentally destructive. According to Scott (1998), "the main solution proposed has been sedentarization and making resource use more 'legible' by confining such use to certain defined areas. National legislation of modern African states tends to favour agriculture, which leaves a visible trace in the landscape as evidence of land use, while pastoral use is more invisible and, therefore, cannot easily be used to justify prioritized access or property rights. Farmers also have the advantage of being present, while pastoralists tend to be absent for parts of the year. Since use rights are granted to those who clear the land and add 'productive value' to it, pastoralists have difficulties policing their rights; farmer-herder conflicts can be seen as the result."

Commercialization of cattle raiding

The emergence of local elites who aim to profit from cattle rustling is a fairly new phenomenon that has changed the scope of the conflict by creating economic incentives that did not previously exist. This has exacerbated the brutality associated with raiding and has created links between the illicit trades in stolen cattle and small arms. Local businessmen and even politicians reportedly fund raids in order to sell cattle on the black market to places as far away as South Africa and Saudi Arabia (Mkutu, 2003). Stolen cattle are also used to supply large towns, which have grown in population through rural-to-urban migration. Whereas small-scale raiding does not deplete entire stocks, commercialized raids with elaborate planning and logistical know-how can render entire communities destitute (*ibid.*). Buchanan-Smith and Lind (2005) suggest that there are typically five large raids in southern Turkana in a given year. Generally, large infrequent raids, coupled with repeated small-scale incidents create an environment of insecurity and financial hardship.

Inadequacy of state security provision

As part of a political campaign that favours sedentary communities over non-sedentary groups, governments have neglected to invest a great deal in infrastructure and public services in the pastoralist border areas, thus exacerbating the lack of state security in the region. Without sufficient roads, accessible lines of communication, and a large qualified security presence, pastoralists have had no choice but to take up arms in order to protect their families and livestock. Moreover, cross-border raiders are immune from prosecution, because governments lack the capacity or infrastructure required to prosecute those involved in acts taking place in other jurisdictions. The Small Arms Survey reports, for example, that nearly 60% of residents living along the Kenya-Sudan border are dissatisfied with security provisions in their communities (McEvoy and Murray, 2008).

In many instances, warriors conducting large commercial raids outnumber security forces. There is also evidence that many local security providers are in collusion with the profiteers of raids (Mkutu, 2003). The lack of state control in the pastoralist region has made way for what has been termed cattle warlords surrounded with armed militia (Osamba, 2000). Without proper security provision, a small number of entrepreneurs will continue to benefit at the expense of a great number of people.

In Kenya and Uganda, where the military's role is restricted to responding to large-scale incidents and carrying out community disarmament programs, governments have armed local defence units to provide security at the local level. These comprise civilians who are given a registered firearm and ammunition without any training or remuneration. In Kenya, for instance, the Kenyan Police Reservists (KPR), armed with Kalashnikov and G3 assault rifles, function as a community task force mandated to respond to local crime and disputes. While the KPR are sometimes effective in defending communities against cattle raids, they are also known to lend out their weapons to warriors for raiding purposes, undermining the very security they are supposed to protect. Moreover, a 2008 ammunition study in Kenya found that the majority of illicit civilian-held ammunition was Kenyan-manufactured, revealing that it had either been stolen from weakly guarded stockpiles or sold by corrupt officials (Bevan, 2008).

Proliferation of small arms

Small arms have been present throughout the region since the early twentieth century. However, past wars in Uganda, Sudan, and Ethiopia left a surplus of weapons in circulation. One event, in particular, that is often recalled is the 1979 raid on the Moroto arms depot in Uganda, following the collapse of the Idi Amin regime. The Matheniko Karimojong sub-clan were successful in acquiring large quantities of weapons and ammunition during the raid (Mkutu, 2007). A similar event took place in Kapoeta, Southern Sudan, in 2002, during the North-South war. After the SPLA captured Kapoeta, which at the time was a military stronghold of the north, security at arms stores was lax or non-existent, and as a result, the Taposa tribesmen living around Kapoeta were able to seize thousands of weapons (Bevan, 2008).

Demographic expansion

High population growth has affected rural as well as urban areas, where farmers increasingly moved onto less productive lands to raise their farms and families. Furthermore, pastoralists such as the Maasai have adopted and/or increased farm cultivation, leading to a loss of pasture and water resources. For instance, in the more arid northern Kenya where agriculture is possible only in isolated islands, population growth in both herds and humans has brought about increased competition with pastoral neighbours for pasture and water, leading to recent armed attacks between the Turkana and Pokot, the Borana and Rendille, the Turkana and Samburu, and the Somali and Boran (Daily Nation, 1999).

The growth of human and livestock populations in East Africa's pastoral areas, although modest, has direct consequences for land management and resource use in these ASAL regions. This is particularly so around permanent water and dry season grazing resources, which are located in more populated highlands and are attracting sedentarizing populations. The concentration of populations directly contributes to economic transformations and political conflict in these regions (*ibid.*).

Conclusions

Natural resource use conflict is a real problem among pastoral communities. It is experienced differently in different places. The understanding of its causes and how it should be addressed is in itself the reason why conflicts persist. It is a problem with value judgments and moral issues rather than being a simple imbalance of power or measurable inadequacy. Realistic judgments must consider the values of those affected by it.

The complexity and context of conflict makes it a problem whose meaning depends on the values and relationships within society. Although it cannot be isolated from those experiencing it, its resolution cannot similarly be focused entirely on them either. There is need to broaden knowledge of this problem: not only of the conditions of the communities in conflict, but also of the whole process of cultural dynamism and social valuation. Such a broad concept of culture plays an important role in understanding conflict.

The interactive governance framework suggested in this chapter provides a new perspective in understanding this problem. For this reason, conflict as a governance issue needs to rely on the collective judgment of various stakeholders through a process that is participatory, communicative and transparent. In addition, this framework presents an opportunity to assess how conflict can be governable. In this case governability can be assessed through the components of the interactive governance, namely systems-to-be-governed, governing systems and governing interactions. Governability assessment would be focused on the systems properties, including diversity, complexity, dynamics and scale.

Table 8.1 presents a list of general research questions which can be used to understand conflict. This list is not exhaustive and each cell can be filled with several questions in a particular context. It is prudent, however, to approach conflict with a lot of flexibility. This is because perceptions people hold on certain things may prove to be different when subjected to closer scrutiny. Since perceptions are powerful and judgmental, an understanding of perception is basic to comprehending conflict.

Table 8.1: An interactive governance framework for understanding conflict in pastoral communities

VARIABLES FOR ASSESSMENT	GOVERNING SYSTEM	SYSTEM-TO-BE-GOVERNED	GOVERNING INTERACTION
Decision making	What governing modes are used to reach decisions? Are they top-down, co-management or bottom-up?	What are the social and ecological impacts of the decisions made to address poverty? Are these decisions considered fair and transparent from the perspective of the poor?	How do individuals, groups and governing institutions interact as they negotiate decisions?
Institutions	How are governing systems institutionalized? How is power exercised, responsibilities and mandates distributed and with what outcomes?	How are social practices, interests and power institutionalized? How do they determine social relationships and the distribution of cost and benefits among stakeholders?	What are the institutional characteristics of governing interactions? What rules exist pertaining to representation, participation and communication?
Value	How do values, norms and principles of governing institutions relate to problem definition, agenda setting and conflict resolution?	What values, principles and norms underpin social action and local institutional formations, decision-making and power relations?	How are values, principles and norms shared among stakeholders in their interactions?

Source: Chuenpagdee and Jentoft (2009).

Perception is a dynamic balance between reality and our mentality. Thus, both the outer reality and inner mentality in this balance must be weighed and the nature of the balance itself must be comprehended (Rummel, 1976). As Rummel defines it, perception is "...a confrontation between the inner mentality and outer reality--a balance between the two. As such, it is a mixed sensual-conceptual awareness of external dispositions and powers. What then we perceive is a situation" and this says a lot about how behaviour is directed. It is also worth noting that the inner mentality is shaped by the socio-cultural context in which an individual belongs, and this is why culture plays an important role in understanding behaviour.

References

Adams, W. and Hulme, D. (2001). "If community conservation is the answer in Africa, what is the Question?" *Oryx*, 35 (3): 193-200.

Adriansen, H.K. (1997). The development of nomadic pastoralism in Africa. A Master Thesis submitted to the Institute of Geography, University of Copenhagen, Copenhagen; pp 105.

Alao, A. (2011). Natural resources and the dynamics of conflicts in West Africa. ECOWAS and the Dynamics of Conflict and Peace-Building; pp 45 – 60.

Archibo, P. (2011). Conflict resolution in Pokot and Turkana. Interview script from 4 October 2011 in Lokiriama.

Bavinck, M., Chuenpagdee, R., Diallo, M., van der Heijden, P., Kooiman, J., Mahon, R., and Williams, S. (2005). Interactive fisheries governance: a guide to better practice. Delft, The Netherlands.

Behnke, R., Scoones, I., and Kerven, C. (1993). Range Ecology at Disequilibrium: New Models of Natural Variability and Pastoral Adaptation in African Savannas, London: Overseas Development Institute.

Benjaminsen, T.A., Maganga, F.P. and Jumanne, M.A. (2009). "The Kilosa killings: Political ecology of a farmer–herder conflict in Tanzania." *Development and Change* 40(3): 423–445.

Bevan, J. (2008). Blowback: Kenya's illicit ammunition problem in Turkana North District. Occasional Paper 22. Geneva: Small Arms Survey.

Birch, I. and Grahn, R. (2007). "Pastoralism – managing multiple stressors and the threat of climate variability and change." Retrieved 18 August 2011, from http://hdr.undp.org/en/reports/global/hdr2007-8/papers/Birch_Isobel%20and%20Grahn_Richard.pdf.

Blench, R.M (2001). 'You can't go home again': Pastoralism in the new millennium. London: Overseas Development Institute.

Bonfiglioli, A.M. (1992). Pastoralists at a crossroads: Survival and development issues in African pastoralism. Nomadic Pastoralists in Africa Project (NOPA) UNICEF, UNSO, Nairobi.

Brockington, D. (1999). "Conservation, displacement, and livelihoods: The consequences of eviction for pastoralists moved from the Mkomazi Game Reserve, Tanzania." Nomadic Peoples, (NS) 3 (2): 74-96.

Buchanan-Smith, M. and Lind, L. (2005). Armed violence and poverty in Northern Kenya - A case study for the Armed Violence and Poverty Initiative. Centre for International Cooperation and Security/ Bradford University, Bradford.

Chuenpagdee, R. and Jentoft, S. (2009). "Governability assessment for fisheries and coastal systems: a reality check." Human Ecology, 37, 109–120.

Daily Nation (1999). "Ethnic tensions worry leaders," July 27, 1999, Nairobi.

DIAL (2012). Conflict resolution and peace building. Available at: http://www.dialafrica.

Desta, S. (2006). "Pastoralism and development in Ethiopia." Economic Focus, 9(3):12-20. Also available at http://dspace.africaportal.org/jspui/bitstream/123456789/34403/1/Economic%20Focus%20%20Vol%209%20No%203_0.pdf?1.

Eaton, D. (2008). Violence, revenge and the history of cattle raiding along the Kenya-Uganda border. Halifax: Dalhousie University.

FAO, (2000). Conflict and Natural Resource Management. Rome, Italy.

Fratkin, E. and College, S. (2004). Ariaal pastoralists of Kenya: Studying pastoralism, drought and development in Africa's arid lands. Boston: Pearson; 160pp.

Fratkin, E. (2001). "East African pastoralism in transition: Maasai, Boran, and Rendille cases." African Studies Review, 44(3): 1-25.

Fratkin, E. and Wu, T.S. (1997). "Maasai and Barabaig pastoralists struggle for land rights in Kenya and Tanzania." *Cultural Survival Quarterly*, 21(3):55-61.

Galaty, J.G. and Johnson, D.L. (eds) (1990). *The world of pastoralism: Herding systems in perspective.* London: The Guilford Press.

GoK (2011). "Kenya National Focal Point on Small Arms and Light Weapons." Retrieved 11 December 2011, from http://www.nscpeace.go.ke/nsc/index.php/knfp.

GoK (2008). "Arid Lands Resource Management Project, Report Turkana." Retrieved 27 June 2011, from http://www.aridland.go.ke/arid_annual_reports/arid_annual_07-08/turkana_07-08.pdf.

Hackel, J. (1999) "Community conservation and the future of Africa's wildlife". *Conservation Biology,* 13 (4): 726-734.

Hesse, C. and Cavanna, S. (2010). *Modern and mobile: The future of livestock production in Africa's dry lands.* London: International Institute for Environment & Development (IIED) and SOS Sahel International UK.

Hesse, C. and MacGregor, J. (2006). Pastoralism: Drylands' invisible asset? IIED Drylands Issue Paper No. 142. IIED, London, UK.

Hogg, R. (1997). Drought contingency planning to support pastoral livelihoods in Ethiopia. A discussion paper prepared for UNDP-EUE, Addis Ababa.

Homewood, K.M. (1995). "Development, demarcation and ecological outcomes in Maasailand." *Africa,* 65:331-350.

Hulme, D. and Murphree, M. (2001). *African wildlife and livelihoods: The promise and romance of community conservation.* Oxford, UK: James Currey.

Hunter, M. (*n.d.*). Nomadic pastoralists - Who are they? Nomadic Peoples Network. Available here: http://nomadicpeoples.net/defining-nomadic-pastoralists/

International Organization for Migration (IOM) (2011). "IOM facilitates inter district peace dialogues between the Turkana and Pokot." *IOM Nairobi Bulletin,* 5(6) pp 1.

Jahnke, H. (1982). *Livestock production systems and livestock development in Tropical Africa.* Kiel: Kieler Wissenschaftsverlag Vauk.

Jentoft, S., and Chuenpagdee, R. (2009). "Fisheries and coastal governance as a wicked problem." *Marine Policy*, 33, 553–560.

Juma, M.K. (2000). Unveiling women as pillars of peace building in communities fractured by conflict in Kenya. An interim report for Management Development and Governance Division, Bureau for Development Policy, UNDP, May.

Kahl, C.H. (2006). *States, scarcity, and civil strife in the developing world.* Princeton, NJ: Princeton University Press.

Kaliamoi, J.Y. (2011). Developments in Alale and surrounding. Interview script from 16 October 2011 in Alale.

Kandagor, D. (2005). Rethinking pastoralism and African development: A case study of the Horn of Africa. www. codesria.org/Links/ conferences/general.../kandagor.pdf.

Karani, I. (2008) Climate change and pastoralism: A case study of Kotido District, Uganda. OXFAM Briefing Paper. Mimeo.

Kerven, C. (1992). Customary commerce: A historical reassessment of pastoral livestock marketing in Africa. ODI, London.

Kooiman, J. and Jentoft, S. (2009). "Meta-governance: Values, norms and principles, and the making of hard choices." *Public Administration*, 87(4):818-836.

Kooiman, J., Bavinck, M., Chuenpagdee, R., Mahon, R. and Pullin, R. (2008). "Interactive governance and governability: An introduction." *The Journal of Transdisciplinary Environmental Studies*, 7 (1):1-11

Kooiman, J. and Chuenpagdee, R. (2005). "Governance and governability." In: *Fish for life: Interactive governance for fisheries*, edited by J. Kooiman, M. Bavinck, S. Jentoft, R. Pullin. Amsterdam: Amsterdam University Press; pp. 325-349.

Kooiman, J. (2005). "Principles for fisheries governance: Introduction." In: *Fish for life: Interactive governance for fisheries*, edited by J. Kooiman, M. Bavinck, S. Jentoft, and R. Pullin. Amsterdam: Amsterdam University Press; pp. 241-244.

Kooiman, J. (2003). *Governing as governance.* London: Sage.

Kirk, M. (1999). "The context for livestock and crop-livestock development in Africa: The evolving role of the State in influencing property rights over grazing resources in Sub-Saharan Africa." In:

Property rights, risk, and livestock development in Africa, edited by N. McCarthy, B. Swallow, M. Kirk and P. Hazell. IFPRI and ILRI, Nairobi.

Knighton, B. (2003). "The state as raider among the Karamojong." *Africa,* 73(3): 427-455.

Lawson, D.W., Borgerhoff, M.M., Ghiselli, M.E., Ngadaya, E., Ngowi, B., Sayoki, G.M., Mfinanga, K.H. and Susan, J. (2014). "Ethnicity and child health in Northern Tanzania: Maasai pastoralists are disadvantaged compared to neighbouring ethnic groups." *PLoS ONE,* 9(10):1-17. e110447. doi:10.1371/journal.pone.0110447.

Lind, J. 2006. "Supporting pastoralist livelihoods in eastern Africa through peace building." *Development,* 49: 111-115.

Little, P.D. (1992). *The elusive granary: Herder, farmer and state in Northern Kenya.* Cambridge: Cambridge University Press.

Locham, R. (2011). European instrument for democracy and human rights. Interview script from 15 October 2011 in Alale.

Lodea, E. (2011). Community developments, role of elders. Interview script from 2 October 2011 in Lokiriama.

Lokuwam, S. (2011). Activities of women and role in community. Interview script from 29 September 2011 in Lokiriama.

Lomor, P. (2011). Livestock trading and security issues in Turkana. Interview script from 1 October 2011 in Lokiriama.

Longolol, E. (2011). Community developments, role of elders. Interview script from 6 October 2011 in Lokiriama.

Lugoe, F. (2011). Aligning and harmonizing the livestock and land policies in Tanzania. ESRF Discussion Paper No. 35. Economic and Social Research Foundation, Dar es Salaam.

Mattee, A.Z. and Shem, M. (2006). Ambivalence and contradiction: A review of the policy environment in Tanzania in relation to pastoralism. IIED Issue Paper No 140, March, 2006. London, International Institute for Environment and Development.

Mburu, N. (1999). "Contemporary banditry in the Horn of Africa: Causes, history, and political implications." *Nordic Journal of African Studies,* 8 (2): 89–107.

McEvoy, C, and Murray, R. (2008). Gauging fear and insecurity: Perspectives on armed violence in Eastern Equatorial and Turkana

North. Small Arms Survey Working Paper 14. Geneva: Small Arms Survey.

Michael, O. (2008). *Conflicts in Ngorongoro District: Causes, consequences and possible solutions.* Oxford: Oxford University Press.

Mieth, F. (2006). *Defying the decline of' pastoralism.* Amsterdam: University of Amsterdam.

Mkutu, K. (2007). "Impact of small arms insecurity on the public health of pastoralists in the Kenya-Uganda border regions." *Crime, Law and Social Change,* 47 (1): 33–56.

Mkutu, K. (2006). "Small arms and light weapons among pastoralist groups in the Kenya-Uganda border area." *African Affairs,* 106/422. p. 47.

Mkutu, K. (2003). Pastoral conflict and small arms: The Kenya-Uganda border region. *Safe World Report;* www.saferworld.org.uk/publications.php/54/pastoral_conflict_and small-arm.

Morton, J., Livingstone, J.K. and Mussa, M. (2007). Legislators and livestock: Pastoralist parliamentary groups in Ethiopia, Kenya and Uganda. IIED, Gatekeeper Series 131.

Mukoo, B.T. (2011). Conflicts in Karamoja. Interview script from 26 September 2011 in Lodwar.

NGONET (2008). A Consultancy Report Commissioned Jointly by ERETO Ngorongoro Pastoralist Project (ERETO-NPP) and the Ngorongoro Local Government and managed by Ngorongoro NGO Network (NGONET). *Arusha Nomadic Peoples,* 34/35: 7–21.

Odhiambo, M.O. (2012). Impact of conflict on pastoral communities' resilience in the Horn of Africa: Case studies from Ethiopia, Kenya and Uganda. FAO Report.

Olenasha, W.T. (2004). Reforming land tenure in Tanzania for whose benefit? A draft report of the Critical Analysis of the Land Acts of 1999. Dar es Salaam.

Olengurumwa, P. (2009). Resource based conflicts in Northern Tanzania: The case of Sonjo and Maasai of Ngorongoro. LLB Dissertation, University of Dar es Salaam.

Omosa, E.K. (2003) *Natural Resource Based Conflicts and Pastoral Livelihoods: A Focus on the Impact of Water Resource Based Conflicts on Pastoralism in Dadajabulla Location of Wajir District, Kenya.* Institute for Development Studies University of Nairobi, Kenya.

Osamba, J. (2000). "The sociology of insecurity: Cattle rustling and banditry in North-Western Kenya." *African Journal of Conflict Resolution,* 1(2): 11–37.

Oxfam (2008). Turning up the heat: Climate change and poverty in Uganda. Kampala: Oxfam GB in Uganda. Available at: www.oxfam.org.uk/publications

Oxfam (2006). Delivering the agenda: Addressing chronic underdevelopment in Kenya's arid lands. Oxfam International Briefing Paper, 88.

Pei, Q. and Zhang, D.D. (2014). "Long-term relationship between climate change and nomadic migration in historical China." *Ecology and Society,* 19(2): 68. http://dx.doi.org/10.5751/ES-06528-190268.

Practical Action (2011). Background information: Commemoration of the Lokiriama Peace Accord to be held on 21st September 2011 in Lokiriama Trading Centre, Loima District Turkana County Kenya.

Raider, P.I. (2011a). Raiding in Pokot and Turkana. Interview script from 23 September 2011 in Lodwar.

Raider, P.I. (2011b). Conflicts between Pokot and Turkana: Perspectives of the youth. Interview script from 19 October 2011 in Alale.

Rummel, R.J. (1991). *The conflict helix: Principles and practices of interpersonal, social, and international conflict and cooperation.* New Brunswick, N.J.: Transaction Publishers. Also available at http://www.hawaii.edu/powerkills/NOTE11.HTM retrieved on 30th November, 2015.

Rummel, R.J. (1976). "Understanding conflict and war. The conflict helix." Vol. 2. Chapter 3. Sage Publication.

Schilling, J., Akuno, M., Scheffran, J. and Weinzierl, T. (2012a). "On raids and relations: Climate change, pastoral conflict and adaptation in Northwestern Kenya (under review)." In: *Climate change and conflict: Where to conflict sensitive climate adaptation in Africa?* Edited by S. Bronkhorst and U. Bob. Durban: Human Sciences Research Council.

Schilling, J., Opiyo, F and Scheffran, J. (2012b). "Raiding pastoral livelihoods: Motives and effects of violent conflict in Northwestern Kenya." *Pastoralism,* 2(25): 1-16.

Scott, J.C. (1998). *Seeing like a state: how certain schemes to improve the human condition have failed.* Yale Agrarian Studies. New Haven (CT): Yale University Press.

Shivji, I. (1993). *Not yet democracy: Reforming land tenure in Tanzania.* London: IIED.

Sikirika, L. (2011). Community developments, role of elders. Interview script from 1 October 2011 in Lokiriama.

Spear, T. (1997). *Mountain farmers: Moral economies of land and agricultural development in Arusha and Meru.* Oxford: James Currey.

Spencer, P. (1998). *The pastoral continuum: The marginalization of tradition in East Africa.* Oxford: Oxford University Press.

Sulle, E. and Nelson, F. (2009). *Biofuels, land access and rural livelihood in Tanzania.* London: IIED.

Talle, A. (1988). *Women at a loss: Changes in Maasai pastoralism and their effects on gender relations.* Stockholm Studies in Social Anthropology, 19. Stockholm: Department of Social Anthropology, Stockholm University, Sweden.

Toulmin, C. (1995) "Tracking through drought: Options for destocking and restocking." In: *Living with uncertainty: New directions in pastoral development in Africa,* edited by I. Scoones. London ITP.

Tulel, I. (2011). Conflicts and developments in Turkana and West Pokot. Interview script from 9 November 2011 in Nairobi.

TUPADO (2011). Turkana Pastoralist Organization Incident Register, 2000-2010.

Umar, A. (ed.) (1994). Symposium for the sustainable development of Northeastern Kenya. Kenya Pastoralist Forum. Nairobi: Crescent of Hope.

UNDP Sudan (2010). Reduction of resource based conflicts among pastoralists and farmers. Accessed at: http://www.sd.undp.org/projects/cp3.htm

UNDP (2011). Drought and potential conflict scenarios in northern Kenya and other arid lands. A Situation Report. Nairobi: United Nations Development Programme.

UNESCO/IUCN (2007). Ngorongoro Conservation Area (United Republic of Tanzania) Report of the WHC/IUCN Reactive Monitoring Mission.

156 PASTORALISM AND CLIMATE CHANGE IN EAST AFRICA

University of Dar es Salaam (UDSM), Centre for Climate Change Studies (2015). Tanzania: Resilience in Semi-arid Economies (PRISE), Research for Climate-Resilient Futures Team at the Centre for Climate Change Studies (CCCS), University of Dar es Salaam.

UN Office for the Coordination of Humanitarian Affairs (UNOCHA) (2011). Humanitarian Report for Eastern Africa. Nairobi: UNOCHA.

UNOCHA (2010a). "Security in Mobility." Retrieved 23 June 2011, from http://ochaonline.un.org/kenya/Advocacy/SecurityinMobility/tabid/6735/language/en-US/Default.aspx.

UNOCHA (2010b). "Pastoralists in the Kenya/Uganda/Ethiopia borders speak on the humanitarian cost of disarmament in their communities." Pastoralist Voices, 1(21): 4.

US Department of State (2011). 2010 Human rights report, Kenya. Retrieved 11 December 2011, from http://www.state.gov/g/drl/rls/hrrpt/2010/af/154352.htm.

USAID (2005). Fact sheet: The Turkana-Pokot-Sabiny cross-border conflict management initiative. November 2005 Issue.

USAID/REDSO, (2003). Greater Horn of Africa Peace Building Project. USAID/REDSO Regional Strategic Objective 6: Enhanced capacity for managing conflict in the region. Technical Report. Available at: http://pdf.usaid.gov/pdf_docs/PNACS438.pdf.

USIP (2007). Natural resources, conflict, and conflict resolution. New York: United States Institute of Peace.

Verstegen, S. (2001). Poverty and conflict. An entitlement perspective. CPN Briefing Paper. Conflict Prevention Network. Berlin

Wepundi, M., Ndung'u, J. and Rynn, S. (2011). Lessons from the frontiers - Civilian disarmament in Kenya and Uganda. Nairobi: Saferworld.

Western, D., Wright, R. and Strum, S. (1994). Natural connections: Perspectives in community-based conservation. Washington, DC and Covelo, CA: Island Press.

Wright, P. (1993). "Ecotourism: Ethics or eco-sell?" Journal of Travel Research, 31(3): 3-9.

Zaal, F. and Dietz, T. (1999). "Of markets, maize and milk: Pastoral commoditization in Kenya." In: "The poor are not us": Poverty and pastoralism in East Africa, edited by D.M. Anderson and Vigdis Broch-Due. Oxford: James Currey; pp. 163-198.

CHAPTER 9

Policies, Legal Frameworks and Institutions Governing Pastoral Systems in East Africa

Adolphine Kateka

Centre for Climate Change Studies, University of Dar es Salaam, Tanzania

Introduction

Pastoralism, as defined by Neely *et al.* (2009), "...is considered the most economically, culturally and socially appropriate strategy for maintaining the well-being of communities in dry-land landscapes, because it is the only one that can simultaneously provide secure livelihoods, conserve ecosystem services, and promote wildlife conservation and honour cultural values and traditions." In East Africa, examples of pastoralists include Kenya's Turkana and Pokot, Uganda's Dodoth and Upe. In the south of Kenya and the north of Tanzania, the Maasai form a well-known pastoralist people. These people, as elsewhere, continue to display wide and deep levels of poverty compounded by adverse policy and by climate change and variability. Pastoralism directly supports up to 20 million people (nearly 10% of the entire population in East Africa), while contributing indirectly to the livelihoods of countless millions through the livestock value chain (Oxfam, 2006; Thornton *et al.*, 2006; Boone *et al.*, 2011; Tenga *et al.*, 2008).

Yet, policymakers in East African countries continue to disregard pastoralism, in particular its contribution to economic growth. As a sector, pastoralism is estimated to be worth USD $800 million a year in

Kenya alone, and over 90% of meat consumed in East Africa and more than 50% of the milk produced comes from pastoral herds (IRIN, 2007). And then there are the environmental values such as nutrient recycling, maintenance of pasture productivity and biodiversity, tree regeneration, maintenance of natural ponds and carbon sequestration for both climate change mitigation and adaptation (Millennium Ecosystem Assessment, 2005). Pastoralists, however, use climatic variability to their advantage to maximize the productivity of their herds. Strategies that include livestock mobility and selective breeding of animals capable of tracking highly dispersed concentrations of nutrients in the rangelands enable pastoralists to make economic use of these lands without degrading the environment (Galvin, 1992; Galvin, et al., 1994).

Both during the colonial and postcolonial eras, the attitude of governments towards pastoralism has ranged from outright hostility to benign neglect. However, there are positive developments. Kenya has taken major steps in recognizing pastoralism in policy statements and documents and has established the Ministry of Northern Kenya and Other Lands intent on supporting pastoralism, while Tanzania's newly created Ministry of Livestock Development and Fisheries has a Department of Pastoral Systems Development. While there are provisions to protect communal lands and customary pastoral land-use practices in Kenya, Tanzania and Uganda, failure to ensure adequate implementation and/ or contradictory legislation has led to substantial alienation of pastoral lands. Eviction of pastoralists from their ancestral lands is increasingly an issue as governments pursue policies promoting foreign investment in commercial agriculture or wildlife-based tourism. The tendency of East African governments to seek to modernize pastoralism has failed to deliver development to pastoral areas as discussed in the following sections.

Tanzania

An overview

This review of policies, laws and regulations governing pastoral systems in Tanzania traces the policies, legal frameworks and institutions through historical lenses from the colonial period to date. The review shows that the impact of colonial policies on the arid and semi-arid areas (ASALs) has endured the passage of time and is still felt today. It is only recently that ASALs have come to the attention of the government and have been recognized in the Tanzania Development Vision of 2025 and MKUKUTA II. In spite of this, recognition of pastoralism as a viable mode of production is not widely supported at policy level. Except for MKUKUTA, there is no specific policy designed for pastoralism in Tanzania. Livestock policies that are in place aim at commercializing the sector. For example, the current vision of the Livestock Policy is that by 2025, the livestock sector will be, to a large extent, commercially run and modern, to ensure food security and improved income for the household and the nation while conserving the environment. The policy does not provide for extensive grazing, which many studies show to be appropriate for the harsh semi-arid environments.

These misconceptions about pastoralism are best illustrated by their place in the National Land Policy adopted by the government in 1996 (MLHUD, 1996). The policy puts pastoral concerns at the periphery of policymaking. From marginalization, ASALs have now come to be seen as lands of potential in terms of tourism, bio-fuel production, mining and hunting, with repercussions for pastoralism and agro-pastoralism. This is understood to be the result of poor understanding of the economic contribution of the ASALs to the national economy.

Tanzania mainland has an area of 945,087 km2 divided into 21 regions. Approximately 70% of the land is rural village land supporting 80% of the population (farmers and pastoralists), 28% is reserved land (forests, national parks, game reserves), and 2% is urban land supporting the rest of the population. The population is currently estimated at 40.00 million people (NBS, 2012).

The ASALs of Tanzania

There is no standard definition of arid and semi-arid lands in Tanzania. Some estimates show that the Tanzanian dry lands are covering between 25% and 75% of the country and others estimate between 45% and 75% (Boesen et al., 1996; Mascarenhas, 1995). The rainfall distribution in Tanzania, as in many places, is used to define ASALs and other zones as shown in Figure 8.1. The semi-arid areas of Tanzania cover the area running from northeast through central to southwest Tanzania with Dodoma as the driest region in Tanzania (Ngana, 1993). In all these areas, the rainfall is highly unreliable (less than 800mm) and mean monthly temperature is above 18 degrees centigrade.

Figure 9.1. Map of rainfall patterns in Tanzania

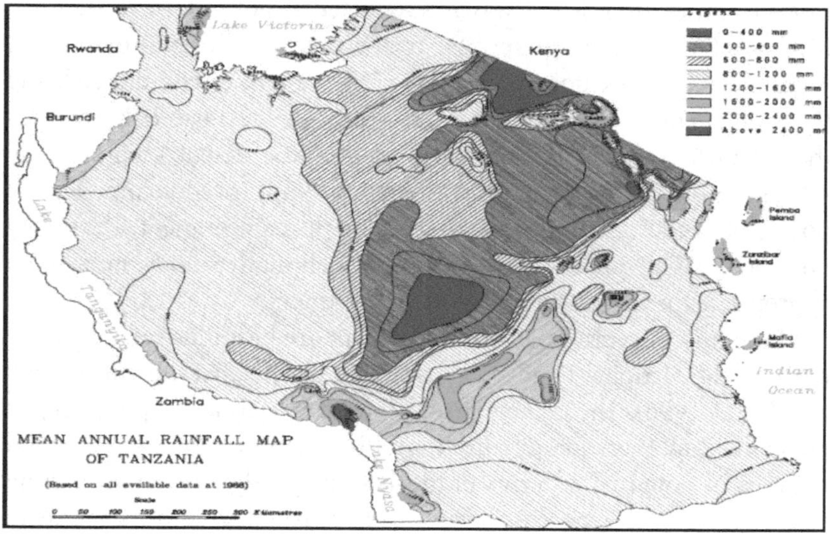

Source: *Shechambo et al. (1999).*

Land use in the ASALs

Land use analysis in Tanzania shows that land in the northern arid lands of Tanzania, which covers about 4000 km2, is nomadic grazing land, game reserves and tourism (LRDC, 1987). The Maasai steppe covers about 8,000 km2 of the country and is exclusive for grazing and game reserve. This area is less suitable for arable farming and potential for irrigation is limited due to the salinity/alkalinity of its soils. The central semi-arid lands and the south-eastern semi-arid lands cover about 15,000 km2. Extensive semi-nomadic grazing as well as small-scale cultivation of drought tolerant roots, cereals and cotton are predominant. There is scope for better integration of livestock in cropping systems, improved livestock production and marketing. However, soil fertility, is generally low (Mwalyosi, 1992).

Pastoral Systems in the ASALs

Pastoral systems may be classified into two broad classes: as ruminants (cattle, sheep and goats) and non-ruminants (represented by pigs and poultry). Ruminants are able to digest rough vegetative matter, *e.g.* pasture, as the main source of feed, while the latter depend on more concentrated feeds, *e.g.* grains. Ruminants are classified under three types of production systems: 1) Specialized and intensive (dairy farms and feedlots); 2) Specialized and extensive (ranches); 3) Agro-pastoralism; and 4) Pastoralism.

Under agro-pastoralism, crops and livestock are raised on the same holding/farm, with crops providing most of the income/subsistence. In pastoralism, most of the livelihood is earned from livestock activities. Agro-pastoralism is predominantly practiced in semi-arid, while pastoralism is predominantly practiced in arid areas. Extensive grazing by the pastoralists is not perceived as a land use to which secure user rights can be claimed. The land on which it is practiced is therefore treated as unoccupied land that can be claimed by more permanent forms of land use (Mwalyosi, 1992).

This problem has become acute, particularly in the last few years due to an increase in instances of outright grabbing of pastoral lands for private ranches as well as encroachment of farm areas into previously marginal grazing areas. Furthermore, opportunity cropping accompanied by soil mining in the most fertile pastoral areas and a significant reduction in the size of pastoral areas available for grazing and watering adds to the problems in this area (Shivji, 2004). This has led to out-migration of many agro-pastoralists from these semiarid areas into Tabora, Mbeya and Morogoro regions. Lack of effective land-use planning, uncontrolled population growth and introduction of Tanzanian village agricultural production system in semi-arid areas have contributed to the present state of affairs (Mwalyosi, 1992).

Economics of the ASALS

The ASALs are home to about 4 million pastoralists and agro-pastoralists with the national herd estimated at 18.5 million (the third largest in Africa after Ethiopia and Sudan), which grazes across 60 million hectares of the 94 million hectares of land resource in the country (only 2% of the national herd is based in ranches, with pastoralists and agro-pastoralists owning the vast majority of livestock) (Sendalo, 2009). It is estimated that livestock is contributing at least 45% to GDP and 30% of agricultural GDP (MALDF, 2015). In addition, Tanzania gets around 90% of its meat and milk products from pastoralism, as well as a vast number of indirect benefits not captured by national statistics.

ASALs also provide significant goods and services, which are marketed nationally and regionally as well as consumed at the household level; these include tourism, hunting, bio-diversity conservation and carbon sequestration credits. The widespread and deeply rooted misconception that ASALs are wastelands, neglects the magnitude of existing economic activity and benefits. However, the value of these goods are poorly captured in national statistics and as a result, government planners frequently make poorly informed choices over investment and policy in the ASALs (TLP, 2012). Contrary to this misconception, dry lands are regions of high economic value (UNIDO, 2012).

Climate Change and Pastoralism

The IPCC Report predicts that climate change over the next century will affect rainfall patterns, river flows and sea levels all over the world. For many parts of the arid regions there is an expected precipitation decrease over the next century of 20% or more (Parry et al., IPCC, 2007). Climate change impacts add to already difficult water management challenges in the ASALs. 'Traditional' approaches to managing water scarcity - based on lifestyle adaptations that minimized consumption and maximized beneficial local use – were dominant up to the 1970s and are still used today in many of these regions.

It is predicted that in the next 10 to 15 years Tanzania will see a continuation of current trends of successive poor rains, an increase in drought-related shocks, and more unpredictable and heavy rainfall events. Beyond this period, the IPCC climate models for East Africa show an increase in temperature of up to 2 to 5°C by the 2080s, with more intense rain predicted to fall during short rains (October to December) over much of northern Tanzania as early the 2020s, and becoming more pronounced by the 2030s. Where this happens, the delicate balance on which pastoral systems depend is undermined as the quality, quantity and spatial distribution of natural pastures are mainly shaped by rainfall (*ibid.*). The economic impacts of these changes in East Africa, including Tanzania, are well discussed in Watkiss *et al.* (2009; 2011).

Predicted changes in rainfall patterns are bound to result in increasingly scarce, scattered and unpredictable pastures. As a result, access to pastures will become more difficult, leading to loss of livestock and livelihoods. The first and most obvious response to drought will be to move animals to areas where there is still pasture and water. Resilient productivity and adaptability to the disequilibrium in ecosystems are a core part of ASALs communities' livelihoods and institutions. Pastoralists have lived within these constraints for centuries and have subsisted on the productivity provided locally and have used their knowledge to devise coping and adaptive strategies to their changing environment and climate variability (Goodin and Northington, 1985).

Most of the major public policies and legal framework lack entry points or are weak to support implementation of priority management options that could enhance pastoralist livelihood and resilience against impacts of climate variability and change (Tenga *et al*, 2008). It is in everybody's interest for pastoralism to succeed, allowing Tanzania to capitalize on the potential of its citizens to manage pastoralism and drive a robust and resilient livestock sector (Njombe and Msanga, 2008).

Policies, Institutions and Laws

The context

Increasing land demand for different uses in pastoral areas and elsewhere triggered a need for regulation and control. Governments responded by adopting several policies, laws and institutions to implement the policies. The most important policies impacting pastoralism include: The Environmental Management Policy, the National Land Policy 1997, the Livestock Policy 2006, and the Wildlife Conservation Policy of 1998 and the National Forest Policy 1998. Enacting laws and establishing institutions to implement them also accompanied adoption of the above policies.

The Environmental Management Policy objectives are supportive of pastoralism as the policy recognizes pastoralism as a source of livelihood that can be sustained in dry areas. The main thrust of the policy and accompanying legislation is to promote restoration and conservation of the natural resource base in pastoral areas. This would be in line with the objectives of a sustainable pastoral system.

The National Land Policy views pastoralism as one of the causes of land degradation and contributing to land ownership and conflict in the country and therefore proposes measures to sendentarize livestock production and control livestock movement. However, the decision to recognize and register communal land tenure and use is a positive development for pastoralism. The decentralization of land-use planning and allocation at the village level will increase participation in land matters by pastoralists in pastoral areas.

Land is a particularly topical issue in Tanzania. The foundation of Tanzania's land legislation dates back to the colonial era, and contemporary land policies are still firmly rooted in the colonial practices. *Ad hoc* commissions were set up to consider applications for land with the task of ensuring either that the land was unoccupied or that compensation would be paid to inhabitants for their loss of land (James, 1971:14). Land without a documented title was deemed to be ownerless Crown Land (Pitblado, 1970:7). Under the British rule, the Land Ordinance was passed in 1923, where German freehold titles were recognised and converted into the English fee simple. Under the ordinance, all other lands were declared to be public lands and were placed under the control of the governor. The governor was empowered to alienate land "in the general interest of the Territory" (Land Ordinance, s.6) by issuing Rights of Occupancy. Rights of Occupancy are akin to government leaseholds and have a maximum term of 99 years. 'Natives' were to hold their land under 'customary law'.

The Wildlife Conservation Policy pursues strategies that antagonise pastoralism in many aspects. For instance, to exclude grazing in Game Controlled Areas is unnecessary after centuries of coexistence between pastoralism and wildlife conservation. Recent studies have demonstrated more compatibility than antagonism between pastoralism and wildlife conservation. But the policy sees only competition. Indeed, even the Wildlife Management Areas (WMAs) are intended to compete with livestock production in natural resources use outside conservation areas. Whereas WMAs, if properly positioned (policy-wise, resource allocation and who benefits) can diversify and increase income in pastoral areas, they can also be a source of resource extraction from the communities. The National Forest Policy 1998 document views pastoralism as being in competition with conservation. But the impact of the policy on pastoralism is not as severe because pastoralism avoids use of forested areas except under severe grazing shortage. Pastoralists avoid forest for fear of contracting animal and human diseases.

The Livestock Policy 2006 gives priority to the modernisation and commercialization of the livestock sector and gives priority to inviting private investors to invest in value adding activities (marketing and processing) and establishment of medium scale ranches on privatized state ranches and smallholder livestock farms on village land set aside for the purpose. The policy promotes public support for controlling major animal diseases in pastoral areas, including vaccinations and dipping. The policy advocates settled livestock farming as opposed to mobility and communal land tenure as strategies for sustainable use of rangelands in ASAL. The stand of the sector ministry on pastoralism is vital for the system to have any chance of acceptance and support in the rest of the government. As discussed elsewhere in the paper, communal land use is the only viable option for pastoral communities, and it is sustainable if the number of dependent families is kept in check. Tanzania's land policies did not change dramatically during the first years of independence. The main thrust has been to transform the countryside from the traditional modes of land use into a modern and monetized economy.

Colonial policies

Pastoral development policy during the colonial period was influenced by the predominant view that pastoralism was inefficient, with low productivity, and led to environmental degradation. Thus, sedentarization was imposed and pastoralists were stripped of their property rights on large portions of rangelands, which were given to the colonial administration for ranching. An important feature of the colonial legacy is that it changed the role and influence of traditional pastoralist institutions, and created national borders which divided well-established pastoral social, economic and ecological units. Therefore, whereas in the pre-colonial period pastoralists moved relatively easily within these units, colonial border demarcation resulted in these traditional movements being re-defined as 'cross-border' movements.

Post-colonial policies

In 1963, the newly independent government adopted the Land Ordinance of 1923, which was crafted by the colonial administration. The tenet of the ordinance was that all land in Tanganyika (now Tanzania) was declared to be public land under the direct control of the British governor in accordance with the Land Ordinance of 1923, Cap 113 (Fimbo, 1992:3). Under the Rights of Occupancy, all freehold lands were converted into government leaseholds by the Freehold Titles (Conversion to Government Leases) Act of 1963, and the later Government Leasehold Act of 1969. This implied that freehold titles were abolished. The following key post-independence policy reforms had significant impacts on pastoralism and the ASALs in general. These are the Ujamaa land reforms in 1974, the liberalization of land policies in the 1980s, and the conservation of rangeland in the 1990s (Shivji, 2004).

Ujamaa land reforms

The Ujamaa Villagization Programme (1974–76) and the Rural Lands Act of 1973 and the Villages and Ujamaa Villages Act of 1975 along with Arusha Declaration of 1967 are identified as being responsible for the plight of pastoralists and many of the land disputes in the ASALs and elsewhere. Nyerere's rural transformation policies provided the rationale for an authoritarian approach to land tenure that was top-down and highly centralized. Land ownership was highly politicized and private accumulation of land was not encouraged (Nyerere, 1962:177). Specifically, the villagization program led to land insecurity and people's loss of control over resources on their lands (Havnevik et al 2000:11). Traditional property rights had no written guarantees, and thus land ownership could be terminated at will by the state (World Bank, 1992:2). In implementing the villagization program, previous land-use systems and property arrangements were disregarded. Furthermore, the Resettlement and Rural Lands Act of 1973 and the Villages and Ujamaa Villages Act of 1975 were not accompanied by any elaborate property rules (URT, 1994a:20). It is viewed that most land disputes are related to suppressed land rights (Tenga, 1997:165). All these reforms led to encroachment on pastoral lands by small landholders (Havnevik et al., 2000).

Planning typically proceeded without consideration of existing land rights and large areas are often set aside for 'communal village projects' or other district or regionally sponsored development projects regardless of whether part of or all the land in question is already used by people in the village. Land used for grazing is commonly considered to be 'unused' and therefore 'available land' in this context (Lerise, 1996; Johansson, 1991; Hoben et al., 1992: 67-80). Village land use is supposed to be regulated by the Village Councils by upholding village by-laws relating to land use, but the villages are not empowered to frame and enforce by-laws on their own initiative without a prior stamp of approval from their District Council and the Prime Minister's Office.

Economic liberalization

From the late 1980s, Tanzania liberalized its economy and implemented a Structural Adjustment Program (SAP), which meant a re-orientation of the communal land tenure policy (Bruce, 1989:5). However, these economic reforms that took place did not address the issue of traditional land rights. Instead, the government launched the National Agricultural Policy, which emphasized village titling through demarcation of village boundaries with an introduction of 99-year rights of occupancy (LTWG, 1992:6). The land survey and demarcation of village boundaries increased the number of boundary disputes between communities in Tanzania. As a result the policy was hardly implemented. The most affected communities in land disputes were pastoralists in the ASALs (URT, 1995:3).

The report recommended a different approach that will protect the land rights of individuals and restructure the tenure system. Thus, in 1995, the Ministry of Lands published the new National Land Policy. The new Land Policy did not change the status quo and the state remained the sole custodian of land (Shivji, 1999:2). The important observation is that the Land Tenure Act No. 22 of 1992 was amended and the policy recognized land as a commodity that can be exchanged. However, recognizing land as a commodity had serious implications for the ASALs. In April 1999, the Land Act and the Village Land Act of 1999 that constituted Land Law Reforms were passed and approved by the Parliament and became operational in 2001. Whereas the amended Land Tenure Act No. 22 of 1992 facilitated the state's administrative control over people's access to land, the Land Act and Village Land Act facilitated the right of citizens to access and own land (Wily, 2000a:2).

In principle, under section 25 of the Village Land Act of 1999, groups of people including pastoralists have rights to access and own land through issuing of a certificate of Customary Right of Occupancy (URT, 1999). The Village Council approves the right and has the prime jurisdiction over all land declared village land. The councils have the sole responsibility of overseeing decisions on the designation of land within the village, using the land-use plan as the land management tool (NLUPC, 1998:14–15). According to the Land and Village Land Act of 1999, land in Tanzania is divided into government land, reserved land, and village land. Village land is further sub-divided into three main categories: (1) communal and public use land; (2) land for individual or family or group use under customary law; and (3) land reserved for future communal or individual use. Although these subdivisions had impacts on all people owning land, the pastoralists were the most affected.

The Tanzania 1977 Constitution, Section 24(2) requires fair and prompt compensation whenever the rights are extinguished for various reasons. The Tanzania Court of Appeal reaffirmed customary tenure rights, thus assuring the protection of the law to thousands of peasants and herders. For the Tanzanian pastoralists, customary tenure right has been difficult to claim and protect, given constant mobility in search of water and pasture. The problem is that often state organs tend to ignore the rights of pastoralists whenever pastoral land is needed for other use, such as commercial farming and/or conservation (Shem, 2010).

However, the two acts remain weak in many fundamental ways. Slow implementation of the laws, particularly village titling and the drawing up of the village land-use plans, is one weakness (Shivji, 1999). In addition, Village Land Councils do not function as a judicial entity, but rather act as a mediator and assist different parties in land disputes (*ibid.*). To address the situation, the government passed a new Courts (Land Dispute Settlements) Act of 2002. The act consists of different bodies at the village, district and state levels, each of which is to deal with land disputes in conflict situations and litigations.

While all these changes in land use were taking place, no sectoral policies were in place to guide government actions in a coordinated manner, until when the first Livestock Policy was introduced in 1983 and in 1998 when the Wildlife Conservation Policy was adopted. For example, the setting aside of several wildlife conservation areas from 1905 to 1978 was done through guidelines, laws and regulations by the Department of Wildlife (MNRT, 1978). By 1981, no policy to guide the development of livestock industry existed. On this issue Mwalimu J.K. Nyerere had this to say: "…We have, for instance, specific zones for crops like cotton, coffee, tobacco and sisal but not like that for cattle. We can have special areas for zebras (national parks) but livestock keepers are hanging" (Mwaikusa, 1997).

Since 2002, some key policies have been formulated which particularly addressed pastoralists in the ASALs. The Grazing-land and Animal Feed Resources Act no.13 of 2010 and the National Livestock Policy of 2012 underscore the importance of pastoralism as a viable mode of production. However, these policies are still inadequate. In particular, there are still knowledge gaps at different levels and among stakeholders on how ASALs and pastoralism benefit the national economy, as well as gaps in policy implementation. It is, however, acknowledged that pastoralism has a lot of opportunities which if exploited will contribute immensely to poverty alleviation in our country (Department for Pastoral Systems Development, Ministry of Livestock and Fisheries Development, Tanzania, 2011).

Tanzania Development Vision 2025

Tanzania's ASALs landscape is changing very rapidly. The Tanzania Development Vision 2025 (TDV) is driving large-scale commercial farming, private investment in natural resource-based activities, tourism, bio-fuels crops production, mining, and using forested areas for carbon credits in the ASALs. According to Vision 2025, Tanzania's aim is to graduate from a least-developed country to a middle-income country. Medium and large-scale investors are to be the engine of transformation. In the 2005 National Strategy for Growth and Poverty Reduction (NSGPR- MKUKUTA), pastoralism is recognized as a "sustainable and viable livelihood." The strategy advocates for the first time for a more "efficient utilization of the rangeland" and for the "empowering of pastoralists to improve livestock productivity".

Local government

Previously, customary pastoralist institutions (usually elders) made key decisions about the use of the ASALs, for example, when and where herds should move to, how to control dry season grazing, how to regulate access to water and so on. With the onset of local and national development strategies, these local institutions have been steadily disempowered as part of the process of change. As well as this alteration, the history of marginalization of pastoralists generally means that the majority of pastoralist citizens are ignorant of the new policies and legislation relevant to their livelihoods (Ellis and Mdoe, 2004). The process of local government reform in Tanzania offers a valuable opportunity for pastoralists to become more engaged in the decision-making process.

Decentralization provides the potential for greater efficiency in the delivery of services tailored to local needs, better management of natural resources and the more active involvement of local people in the management of their affairs. However, pastoralists are not sufficiently involved in local government planning processes; village and district level plans often do not benefit from, or support, their productive strategies. A compounding issue is that pastoralist mobility may hinder their participation in mainstream political processes at the local level. In mixed ethnic communities, there is sometimes the perception that pastoralist presence is transient in nature. This leads to local political processes being monopolized by the more sedentary communities in these areas.

That the land policies of this period are ridden with inconsistencies and contradictions is the product of conflicting policy signals, multiple agencies of implementation and vague and legally flawed land legislation. Three distinct institutional channels operated through the regional and district levels, each with their own vertical chains of command emanating from their national head offices. First, the RIDEPs were run by their respective donor agencies and were coordinated by the PMO, which was also in charge of planning. Second, the government oversaw the activities at each level and implemented its policies through its appointed executive officers. And third, the party machine policed the implementation of the party directives and 'guided' the various activities (See Belshaw [1982:295] for a schematic presentation of the complex administrative structure for 'regional planning' after decentralization). The multiplicity of policy objectives and implementers/administrators precluded the emergence of a more coherent policy vision and contributed to the often-haphazard implementation of the stated policies (Kleemeier, 1984:190-92).

From the colonial times, the pastoral policy in Tanzania has been largely influenced by the modernization paradigm viewing pastoralism as unproductive and degrading the environment (Mattee and Shem, 2006; Ndagala, 1990), whereby livestock keeping needs to be brought in line with modern development through enclosure of the commons and the establishment of permanent settlements.

On the one hand, government policies have been encouraging farmers to enlarge their areas to make the country self-sufficient in food as well as in cash crops. On the other hand, pastoralists have been told to reduce the number of livestock, which are more productive (Benjaminsen et al., 2009). However, the requirement for keeping exotic breeds without necessary infrastructure and given inadequate extension services (Benjaminsen et al., 2009), is more risky than allowing pastoralists to continue with their indigenous breeds, which are relatively resilient to harsh environment and diseases (URT, 2006).

Neoliberal policies and reforms introduced in the 1990s had significant impacts on pastoral communities since government support has been reduced. Support provided by the government before neo liberalization included dips for control of ticks and tick-borne diseases and dams for watering livestock. Such costs today are incurred by pastoralists themselves and partly under cost-sharing arrangements by beneficiaries (Benjaminsen et al., 2009).

The National Land Policy (1995), Land Act (1999) and Village Land Act (1999) all reinforce the idea of enclosing and registering village land for privatization. This perspective is considered as a positive approach in the development of farming communities, but as an obstacle to pastoralists who need to be flexible searching for pastures; in this particular context they are virtually marginalized.

According to URT (2005), "livestock policy is informed by the modernization perspectives encouraging development of a commercially oriented, efficient and internationally competitive livestock industry" while the communal and mobile system of exploiting resources of pastoralists is criticized on the ground.

Generally, because of the divergent notions of pastoralism held by the policy makers, this economic activity is undervalued and neglected. If pastoralism is to be valued and supported, then infrastructure and extension services must be put in place to ensure that the vision of the Livestock Policy (2006) is achieved. The vision states that, "By year 2025, there should be a livestock sector, which to a large extent shall be commercially run, modern and sustainable, using improved and highly productive livestock to ensure food security, improved income for the household and the nation while conserving the environment."

The mission of the livestock industry is: "To ensure that livestock resource is developed and managed sustainably for economic growth and improved human livelihoods" URT (2001); despite the 'good policy statement' of the policy as it is clearly stated, very little has been done to reach that dream.

The livestock sector is still improperly managed and the expected wealth from it remains an unfulfilled desire. A common belief even among the political leaders themselves is that although there are valid driving forces to improve the sector, pastoralists do more harm to the overall economy than good due to continuous mobility, haphazard mobility fuelling conflicts with agro pastoralists (particularly farmers) and making disease control difficult (URT, 2001; Benjaminsen et al., 2009). For instance, in a speech in 2006, the President of the United Republic of Tanzania touched the issue of livestock: "We are producing little milk, export very little beef, and our livestock keepers roam throughout the country with their animals in search for grazing grounds. We have to do away with archaic ways of livestock farming. I, therefore, create a separate ministry for livestock" (Hon Jakaya M. Kikwete, URT President in a press conference announcing his cabinet on 4th Jan. 2006).

This statement by the former president, which in most ways affects the policies, shows how the president views narrowly pastoralism and its dynamics. Policies dispossessing pastoralists of their land – especially their best rangelands on which ability to mitigate risk and enhance resilience depends – and converting it to conservation or irrigated, often commercial, agriculture are perpetuating a vicious circle of increasing poverty, resource conflict and environmental degradation that reinforces the very preconceptions and misunderstandings surrounding pastoralism as a livelihood system (Dyer, 2008). For instance, when poorly applied, village-based land-use planning approaches such as that implemented in Tanzania through the Village Land Act can also create artificial borders (IIED and SOS Sahel, 2010).

The persistence of such misunderstandings is paradoxical given the wealth of research that exists demonstrating the ecological, and to a lesser extent, the economic viability of pastoralism in environments characterized by scarce and erratic rainfall. The inability of pastoralists themselves to articulate the rationale of their livelihood system, the scope and scale of its values and to make economic arguments for pastoralism further exacerbates their marginalization (Dyer, 2008).

This is not evidence-based, but it is the practice. Not only is there no consensus on the appropriate economic model of pastoralism, likewise no mechanisms exist to inform government decision makers of the comparative advantages of pastoralism over alternative land uses. Even when policies are broadly supportive of pastoralism, they still advocate for increasing its productivity through modernization.

These negative preconceptions and perceptions result in minimal government investment in support of pastoral development with parallel efforts either to modernize it or to replace it by expropriating the rangeland and other natural resources on which it depends for other uses. Likewise, the alienation of pastoralists from their land, and the restriction of their mobility is one aspect of a larger crisis affecting pastoralism in Africa and Tanzania in particular.

Livestock mobility is the means by which pastoralists access poor and unevenly distributed resources in rangeland areas, while ensuring their livestock graze off the most nutritious pastures available. However, the legislative environment governing pastoralism has historically been anti-pastoralist in seeking to restrict mobility and valuing other forms of land use (such as agriculture) more highly than grazing of animals. Legislative instruments have been used by most independent African states and politicians to legitimize alienation of pastoral land as well as marginalization of pastoralists.

Regional policies

Apparently, pastoralists and the ASALs in general are not only affected by national policies but also by regional policies developed by regional organs such as the:

- East African Community (EAC),
- Intergovernmental Authority on Development (IGAD), and the
- African Union (AU).

These bodies develop policy frameworks that, when adopted by national governments, open up ASALs for development. Such policies include the:

1. East African Protocol on Environment and Natural Resource Management,
2. IGAD Livestock Policy Initiative,
3. AU Policy Framework for Pastoralism in Africa,
4. AU Framework and Guidelines on Land Policy in Africa, and the
5. African Convention on Conservation of Nature and Natural Resources.

These regional policies strongly promote ASALs development, pastoralism and conservation. Although most of the regional frameworks have been developed through intensive consultative processes, the fear is that many of these policy frameworks may be met with non-implementation. Worrying also is the fact that the protocols, agreements and frameworks lack a resource allocation pathway to implement. Within the AU, the framework is driven by recognition that reduction of pastoral poverty is central to the achievement of the Millennium Development Goals (MDGs), because pastoralists represent a substantial segment of the population in many African countries.

The framework suggests that pastoral development efforts must go beyond single-sector technical approaches, and embrace indigenous knowledge, innovations for sustainable natural resource management, effective governance, and further integration of pastoral livelihoods with expanding market opportunities. The big challenge is that the regional nature and the provisions in these frameworks mean that their implementation will require comprehensive legislative, institutional and operational measures at national level to achieve coordinated, decentralized, transparent, efficient and cost effective delivery of services in biodiversity conservation and ASALs development.

Adopting regional policies thus has far reaching implications at national level. It calls for implementation but also for improved governance by reviewing and restructuring existing institutional settings. It also calls for creation of entirely new institutional arrangements at local, national, regional and global levels. Equally important is the need for harmonization of the current national policy environment. This is because current national policies relevant to ASALs have been developed from sectoral perspectives, with conflicting regulation as a result. Implementing supra national policies requires further review and policy harmonization.

Conclusion

Although within the TDV 2025 and MKUKUTA, the Tanzania overall policy environment is now broadly favourable to pastoralism, many of Tanzania's specific development plans and sector policies are far more ambivalent. Many policies promote a modernization agenda, with pastoral land demarcation, titling and the elimination of mobility being key proposals to increase the productivity of the sector. While these may appear to make sense, such approaches are not adapted to rearing livestock in the dry rangelands of Tanzania, as confirmed by research from other ASALs in countries such as Kenya, Ethiopia, and Botswana. Through livestock mobility, and negotiating access to resources managed under common property regimes, pastoralists are able to ensure higher productivity per hectare than under ranching under similar dry-land conditions. Furthermore pastoralism – if properly supported – minimizes disease transfer, avoids the overexploitation of natural pasture and water resources, and reduces risks from coming into conflict with other land users.

In recognition of the vulnerability and potential of ASALs, the Tanzanian government has set up a Department of Pastoral Systems Development that has been operating from the Ministry of Livestock Development and Fisheries since 2006. The National Livestock Policy of 2006 and the Grazing-land and Animal Feed Resources Act no. 13 of 2010 underscore the contribution of pastoralism as a viable way of life. Focusing on eradicating pastoralism, many governments continue to underestimate its economic contribution to their economies.

Kenya

Background

Kenya is home to an estimated four million pastoralists, constituting more than 10% of the population. Pastoralism is the main source of livelihood for communities living in arid and semi-arid lands (ASALs). Pastoralism makes a significant contribution to gross domestic product (GDP) of around 10% in Kenya; it provides the majority of meat consumed in the country and accounts for 70% of the meat produced in Kenya; and it provides a livelihood for tens of millions of people who live there (GoK, 2007). Pastoralists are the custodians of dry land environments, providing services through good rangeland management. ASAL areas cover about 80% of the land area of the Republic of Kenya (OP, 2011). The pastoral human population is estimated at 13 million, accounting for about a quarter of the national population. It accounts for 90% of employment and 95% of family income in ASALs (GoK, 2010). Pastoralism in Kenya is based on cattle, goats, sheep and camels. On several fronts, pastoralism is more important in Kenya than it is in any of the other East African countries (Burundi, Kenya, Rwanda, Tanzania and Uganda).

Despite providing such value, pastoralist areas in East African countries tend to have the highest incidence of poverty and the least access to basic services compared with other areas. Pastoralists face a number of challenges that hinder their way of life and stifle their ability to adapt to changes in their external environment. Taken together, these challenges account for the poverty and lack of essential services. They can be grouped into four main categories: climate change, political and economic marginalization, inappropriate development policies, and increasing resource competition.

Pastoralism, Governance and Climate Change

Pastoralists in Kenya, as in the rest of East Africa, are learning to live with climate change and variability and adapting to its impacts accordingly. Over the next 15 years, droughts and poor rain will be the features of these arid and semi-arid lands. An intergovernmental panel on climate change models for East Africa show an increase in temperature of around 2-4 degrees by 2080 with more intense rain predicted to fall in the short rains over much of Kenya by 2020. Significant negative consequences include loss of livestock through heat stress, loss of land to agricultural encroachment as the rise in rainfall raises the productive potential of arid areas, an increase in frequency of flooding, and the spread of human and livestock diseases that thrive during the wet season.

Political and Economic Marginalization

Pastoralists in Kenya, as in Tanzania and Uganda, have been marginalized in decision-making processes. As a result, underinvestment and social and economic vulnerability are common features in these communities. Marginalization is generally based on their geographical remoteness, ethnicity, and livelihood, which is seen at policy level as outmoded and needs to be replaced by modern livelihood systems. Furthermore, pastoralists are not aware of their rights and have no experience of government accountability; they have thus been unable to defend their rights. Because of poor infrastructure, there is limited private investment in the arid areas of Kenya.

Past Government Policies on Pastoralism

The initial colonial government policy on livestock was directed at controlling the then rampant animal diseases in order to reduce livestock losses in the country herd and to protect settler farmer/rancher stock. Concurrently, the policy encouraged the alienation of large chunks of high potential rangeland in the central and northern Rift Valley highlands, around Mount Kenya and in the eastern provinces to large scale ranching by white ranchers, while pastoralists were assigned the lower potential rangelands in the southern part of the Rift Valley (Maasailand) and the arid northern and eastern dry lands.

The settler farmers were facilitated to establish commercial ranches through generous land grants, often under freehold or long-term lease tenure on land from which indigenous herders were removed. For example, about 17,200 pastoralists with about 2 million livestock were moved from the high potential central Rift Valley Lakipia to make room for 48 white farmers; the pastoralists were then moved again to Kajiado and Narok around 1901 (Langat, 1986).

For most of the 20th Century, rangeland management copied from the temperate grassland and stable conditions of North America (Morton *et al.,* 2007). Development focused on sedentarization and the assignment of fixed grazing lands, denying pastoralists their traditional land rights. The model failed in Kenya's harsh and variable arid lands, highlighting the critical importance of mobility in dry land resource management (Scoones, 2007). In Wajir Kenya, the sedentarization approach has led to reduction in wet season grazing land, leading to fewer areas of fresh pasture following the rains, while the areas grazed in the dry season get no chance to recover. Drought reserve areas have disappeared. Dry or wet season grazing areas no longer exist due to the proliferation of settlements and as a result of constrained mobility.

Pastoralists were organized into a number of grazing schemes to be developed as "models" of proper land use, particularly in livestock management and range resource utilization (Ng'ethe, 1993). The government provided the basic livestock management infrastructure and services; each participating pastoralist was allotted a livestock quota to be kept on the scheme and committed to sell any "surplus" stock over a specified period. Livestock numbers increased rapidly beyond the holding capacity of the schemes, for pastoralists preferred to leave the scheme instead of honouring their commitment to sell surplus stock! By the end of the 1960s, the schemes were judged failures and the land was turned into individual ranches (Langat, 1986).

After independence, the government gave more attention to financing intensive agriculture in the high potential zones, with minimum investment in the pastoral areas, as that would have resulted in low return on investment. It was not until the late sixties that government accepted recommendations that dry lands had adequate potential to justify public support (Veit, 2011). The Department of Range Management was created within the Ministry of Livestock Development, later to be moved to the President's Office (Belshaw, 1982). The division obtained technical and financial support from FAO and UNDP.

A major investment for the development of rangelands (The Kenya Livestock Development Project (KLDP) I, II and III) was launched in 1969 to support the meat subsector development, including group ranches in pastoral areas, grazing blocks in the east rangelands, individual commercial ranches, improvement of livestock marketing infrastructure (stock routes and holding grounds) and marketing services under the newly established Livestock Marketing Division in the Ministry of Agriculture.

Finally, the Kenya Meat Commission was rehabilitated and expanded to take in the anticipated inflow of slaughter livestock. Finances provided by a consortium of external financiers, including the World Bank, USAID, CIDA (Canada), and ODA through the Agricultural Finance Corporation and the Co-operative Bank supported the implementation of the above. The above activities demonstrate the commercialization policy pursued by government in the 1970s and early 1980s. An evaluation of the impact of the investment concluded that despite the substantial disbursement done, there was very little impact on pastoralism (MOARD, 2001; MLD, 2008).

However, although some planned objectives were missed, the program has left Kenya with a vibrant commercial meat sub sector consisting of commercial ranches with the best beef breeding stock in East Africa, modernized processing and marketing infrastructure and services and a vibrant meat trading sub sector, attracting slaughter animals from all the surrounding countries to supplement national supply in meeting internal and export markets.

The development of group ranches formed the most important component impacting pastoralists. Of all the 159 group ranches that were established, only 60 were included in the KLDP (Ng'ethe, 2008). The main objectives for assisting group ranches were (ibid):

- Prevent encroachment into pastoral territories,
- Promote efficient use of rangelands,

- Stimulate investment in rangeland development, and
- Encourage pastoralists to market a larger percentage of their animals.

It was hoped that, overall, the result would be a greater off take of animals and fewer livestock to counter the prevailing trend of overstocking. Some positive results undoubtedly came out of the development of the group ranches for the Maasai, including:

- Individual appropriation of land was scaled down and the influx of non-Maasai into Maasai land was stemmed;
- Development projects funded the construction of some livestock production infrastructure, such as boreholes, dams, troughs, pipelines, and cattle dips; and
- Schools, shops, and health clinics were also established.

Negative results commonly noted include the following:

- Poor project implementation,
- Land degradation following members' refusal to honour their commitment not to hold more stock beyond the holding capacity of the ranches,
- Reduction in the power of customary authorities without the development of an appropriate substitute,
- Divergence between the boundaries of group ranches, customary land management units, and ecological units, and
- Elimination of customary property rights without the development of appropriate substitutes (Kituyi and Kipuri, 1991).

Nonetheless, despite the heavy investment, stocking rates, herd mobility, and marketing behaviour did not change significantly. Indeed, Homewood (1993) reported that after 20 years of group ranches, there are no significant differences in livestock production, wealth, or human nutrition between the Maasai in Kenya and their counterparts in Tanzania.

As problems in managing group ranches mounted, pressure for dividing the ranches increased mainly due to (a) a strong desire by members to own land title deeds for use as collateral for loans, (b) unregistered young men who became adults pressed to be registered as ranch members in their own right, and (c) many ranch members were frustrated by inefficiencies of management committees (Veit, 2011). After initial resistance, government sanctioned subdivision if certain conditions were fulfilled, including settling outstanding ranch loans. Most ranches are in various stages of subdivision.

The failure of group ranches in Kenya is regrettable as it was a well-researched and pretested initiative to reform collective land tenure in harmony with commercialization and sustainable natural resource use, a challenge facing the future of pastoralism in many African countries. Subdivision does not appear to provide the answer to the pastoral problem of poverty, marginalization and degradation of natural resource based pastoral areas. Instead, subdivision of group ranches lead to small land allotments, increased cultivation, and increased land sales that lead to landlessness for many. It also led to more cultivation on fragile marginal lands, constraining integrating wildlife in pastoral areas (Veit, 2011).

Current Policy and Legislation and Institutions on Pastoralism

There is renewed interest in giving priority to the development of pastoral areas in order to reduce the gap that exists between pastoral and non-pastoral areas in terms of development as well as exploiting the potential that exists in the ASALs. The government has drawn the National Policy for the Sustainable Development of Northern Kenya and other Arid Lands to guide and coordinate the national efforts in the development of ASALs in which pastoralism forms an important component (PO, 2011). The goal of this policy is to facilitate and fast-track sustainable development in Northern Kenya and other arid lands by increasing investment in the region and by ensuring that the use of these resources is fully reconciled with the realities of people's lives. The most salient points in the policy document are as follows:

- *The Goal*: To improve the standard of living of communities in the ASALs and ensure sustainable livelihoods.
- *Mobility* is recognized as a rational and sophisticated response to environmental conditions. However, mobility in many pastoral areas of Kenya is curtailed by settlements, administrative boundaries, conflict, and land alienation.
- *Pastoralism* is recognized, through legislation, as *a legitimate form of productive land use and development*.
- Develop *participatory land-use plans*, which rationalize the location of settlements in arid areas.
- Repeal the Land (Group Representatives) Act and provide for pastoralism in the Land Act and institute alternative methods of registration that define individual rights in pastoral communities while *allowing pastoralists to maintain their unique land-use system and livelihoods*.

- Ensure that the land rights of women in pastoral areas are recognized and protected.
- Provide for flexible and negotiated cross-boundary access to protected areas, water, pastures and salt licks among different stakeholders for mutual benefit to facilitate the nomadic nature of pastoralism.
- Ensure that all land uses and practices under pastoral tenure conform to the principles of sustainable resource management in non-equilibrium environments.
- Review the boundaries of reserved lands in pastoral areas to determine the current need.

Based on the policy document, Kenya is taking pastoralism to a higher level of recognition as a legitimate source of livelihood than are other countries in the region. The results of recent socio-economic studies have been incorporated in the policy. This includes recognizing livestock mobility and communal land use and accumulation of herds as essential strategies in pastoral areas.

Legal and Institutional Frameworks

The 2010 Kenya Constitution provides for tenure and registration of community land (GOK, 2010). Article 63(4) specifies, "...Community land shall not be disposed of or otherwise used except in terms of legislation specifying the nature and extent of the rights of members of each community individually and collectively." Article 63(5) requires parliament to pass legislation to implement the group ranch provisions. There are many laws that govern the land, natural resource management, provision of services and regulation of the sale of livestock and produce. Apart from the multiplicity of laws that impact on pastoralism, administration of the laws has been adversely affected by insufficiency and inaccuracy in record keeping.

The main institutions impacting pastoralism are the Kenya Government, the private sector and non-governmental organizations. The GoK has demonstrated resolve to prioritize ASALs in terms of investment, governance and security (OP, 2011). The main government agency for implementing government policy on pastoralism is the Range Management Division, recently transferred from the Ministry of Livestock Development to the Office of the Prime Minister. The relocation of the division emphasizes the multi-sectoral feature of pastoral development and the importance of pastoralism in the dry lands. The division officers are at county, district and lower levels of government, reflecting the decentralization policy of the country. The participation of all stakeholders, especially the pastoral community representatives at all levels of decision making is emphasized in the policy document. Effective participation of pastoralists is made easier in Kenya, given the powers local government have over government budget under the new constitution and the dominance of pastoralists in ASAL districts.

Kenya has a more vibrant private sector than other countries in the region. The livestock trade involving purchase of live animals, processing of meat and marketing of meat internally, and export of the rest is reasonably well organized. Government policy is to promote adding value on pastoral products in partnership with the private sector. The policy document stresses the role of private sector in supplying inputs and procuring pastoral produce for internal markets and export. NGOs are encouraged to assist pastoralists to increase their capacity to add value on pastoral products in order to increase income and diversify sources of livelihood, as well as build their knowledge and skills in negotiation and advocacy for their rights.

Conclusion

Kenya is highly vulnerable to climate change because it is a predominantly dry country. The policies that were designed during the colonial period met little success. However, the independent Kenyan Government introduced further policies that aimed at improving the conditions of the pastoralists. It is currently implementing the Kenya Arid Lands and Resource Management Project (ALRMP), which addresses climate change issues. Kenya Adaptation to Climate Change in Arid Lands (KACCAL) is a project embedded within the highly successful ALRMP. The objectives are to improve national coordination of information and action for management of climate risk, integrate a long-term climate risk perspective in local planning and investments, support community driven initiative on livelihoods and provide technical assistance at the local level (World Bank, 2007). Importantly, the current decentralization policy is having a positive impact on livestock keeping in Kenya. It emphasizes the multi-sectoral nature of pastoralism development and the importance of pastoralism in the dry lands.

Uganda

Background

The Republic of Uganda lies on the equator, between latitudes 4° 12' North and l° 29' South as well as longitude 29° 34' East and 35° East on the East African plateau. The country is land locked and bordered by Sudan to the north, Kenya to the east, Tanzania to the south, the Democratic Republic of Congo to the west and Rwanda to the southwest. The total land area is 241,038 sq. km. The mean annual maximum temperatures range between 18-35°C; while the mean annual minimum temperature range is 8-23°C. The rainfall ranges between 650mm and 2,000mm per year. The driest area of the country is in the north-eastern part, inhabited by the semi-nomadic Karimojong tribe. Most of this population is supported by agriculture consisting of crop, livestock and fisheries production.

In Uganda, over 95% of the indigenous animals in the country are locally adapted and indigenous breeds owned by resource-poor farmers who are mainly smallholders and pastoralists. These animals remain the major contributors to food in the country despite past and present importation of exotic germplasm to improve the livestock production. The increasing demand for exotic breeds affects efforts of conserving and developing the locally adapted and indigenous ones. Equally, as a result of changing trends in demand for livestock products, it's evident that the resource-poor rural farmers who own the majority of the locally adapted and indigenous breeds are now seeking better sources of income by crossbreeding their stock with higher producing recently introduced breeds.

Initiatives for the sustainable conservation of local breeds in Uganda are reflected within the legal, policy and intuitional framework of the country. However, the operational and funding resources required to promote conservation activities at the national level are in short supply. Government policies and legislation, particularly the Animal Breeding Act (GoU, 2001) provides for public sector intervention related to the conservation and utilization of local breeds in the country.

Livestock Sector

In Uganda, pastoralists constitute 22% of the population, *i.e.* around 5.3 million people. The livestock sector is dominated by mixed farming smallholders and pastoralists. It is this group of farmers who produce the bulk of the milk and meat for commercial as well as for domestic consumption (Mwebaze, 1999). During the year 2002, for example, the livestock sector, which is dominated by cattle in value, produced an estimated 1,000 million litres of milk and 107,000 metric tons of beef. Nevertheless, the annual per capita availability of meat and milk remains low, at only 6kg of meat and 40lt of milk, respectively, compared to the FAO/WHO recommended 50kg of meat and 200lt of milk per annum (FAO, 2002). All of these notwithstanding, the livestock sector make a substantial contribution to the overall economy of Uganda. For instance, in 1997 the sector contributed a significant 21.7% of the monetary value to the agricultural component (MFP and ED, 1998). Overall, pastoralism is not as important to Uganda as it is to Kenya and Tanzania, as the sector contributes only 7.5% of Uganda's GDP. Extensive livestock production is important in only 25% of the country (Oxfam, 2008).

Generally, the livestock production systems are characterized by little or no modification of the production environment resulting in low rates of survival, reproduction and effectively low output. The systems account for the production of most of the locally adapted breeds of all species important for food and agriculture in Uganda. The breeds arc kept mostly by farmers with peasantry and subsistence backgrounds. Some 'elite' farmers still use this production system due to cultural dictates. The traditional pastoralists of western and north-eastern Uganda, for example, keep cattle for prestige and numbers, with no or little regard to individual herd productivity.

In the medium input system, there is an increased human intervention in the form of input necessary for improved productivity and general management of animals. The primary production in this system is semi-intensive (Mugerwa, 2001). It includes different sizes of padlock housing during part of the day (especially at night) and supplementation in feeding. The system is used mainly for production of recently introduced breeds of cattle, such as the Friesians. In the high input system, farmers ensure high levels of animal survival, reproduction and output. This is the production system used for recently introduced and continually imported breeds.

Extensive livestock herding, which is common with livestock keepers, is carried out on communal land in the so-called "cattle corridor", which forms a crescent from the southwest of the country to the northeast. Following privatization of the commons, most of the grazing areas in the southwest and central districts have been segmented into about 165 fenced individual ranches. This has contributed to displacement and landlessness, with some landless herders moving to the west and southern parts of the country.

Colonial Policies on Pastoralism (1900 - 1962)

"Since the colonial times," according to Ococh (2013), "livestock improvement policies in Uganda have been biased towards milk and beef production for commercial purposes using mainly exotic germplasm." Policy towards the livestock industry in Uganda was based on three main assumptions (ICPALD, 2013), namely:

- The climate was too harsh for exotic cattle to survive,
- Disease control particularly tick borne diseases was not envisaged, and
- No change foreseeable in the traditional cattle keeping habits.

"Research was initiated to improve the cattle stock based on selective breeding within the indigenous animals for resistance to tick borne diseases and increased milk yield to ensure that the calf got a good start in life and not to breed up a purely dairy type of animal. Selection on the basis of conformation for beef production and stamina for work were secondary objectives." It was soon realized that selective breeding of the indigenous stock would lead to very slow genetic progression which would not bring about the desired production levels in cattle. The policy thereafter changed in favour of importation of exotic genes. Therefore, in 1960, the institution referred to as the Animal Breeding Centre in Entebbe (ABC) was established to provide artificial insemination services.

Post-independence Era (1962 to 1972)

According to Ococh (2013) the Kenyan Government policy on livestock development in this "...era continued to lay emphasis on cattle for both milk and meat." The emphasis was put on promoting transition to modern animal husbandry, encouraging the development of commercial farms in the extensive areas of the unoccupied lands and the establishment of Government and parastatal ranches and dairy farms to breed stock for private ranches in Ankole-Masaka, Buruli, Bunyoro and Singo ranching schemes (UG, 1966).

After 1987 the government issued the Rehabilitation and Development Plan, 1987-1991, which had as a prime objective the restoration of productive capacity of the economy in key sectors. Priority areas for the animal industry emphasized disease control and increased animal production. Following the positive effects of development in 1992 the government outlined important objectives in the livestock sector which included among others: increased self-sufficiency in meat and milk products through encouragement of improved livestock production, the rehabilitation of extension and research services and liberalization of pricing and marketing policies. The ranching schemes in the country were initially supported by government for purposes of commercial livestock production. By way of the Ranch Restructuring Policy instituted in the early 1990s, landless cattle keepers have had to resettle or move out of the area (Behnke, 2010).

The implication has been for the displaced pastoralists moving southwards into Tanzania, or northwards into other districts of Uganda. Most of the communal land in the southwest and central corridor was individualised into commercial ranches, currently numbering 167 (*ibid.*). No change in land tenure/use occurred in the north-eastern part of the cattle corridor, partly because the latter area is more arid. An evaluation on the impact of individualization/ privatization of pastoral land indicated it increased investment in improved technologies; increased wealth to investors; decreased ownership disputes but increased risks of droughts. Pastoralism has been squeezed out of the south, central and west of the country. It remains important only in the Karamoja District and other dry areas of the northern districts.

Current Policies, Legislation and Institutions

Pastoralism in Uganda is regulated through a number of laws. Some laws are general but they impact on pastoralism. First and foremost, the Constitution of Uganda vests land stewardship with the citizens of Uganda. The Land Act of 1968 categorizes land into four categories: (a) customary, (b) individualized, (c) state, and (d) leasehold. Pastoralism is practiced under category (a). Under this tenure system, land is held, used and disposed of following the customary regulations of the concerned community (Mugerwa, 2001). In its pure form, it has an in-built mechanism for conflict resolution and control and access to grazing. But its power has been eroded by outside factors. In many parts of the cattle keeping communities, elite livestock owners and crop farmers have turned the cattle corridor system into individualized or privatized land use. The expansion of state land, which includes land set aside for conservation, has also greatly reduced grazing land all along the cattle corridor.

Currently government policy promotes commercialization, including private and public investment in livestock marketing and processing infrastructure and services. The mission of the government is to have an export abattoir, which is about to be launched, financed by private capital and equity funds from herders through their cooperatives. Several others are planned. The National Animal Breeding Policy formulated under the auspices of the DANIDA supported Cattle Breeding Project has the goal of increasing income to stakeholders, including herders, and meeting the demands of local and external markets. In order to mobilise the internal and external resources needed, the policy promotes public-private partnerships, involving private investors, government and herders through farmer associations.

The policy, along with its Action Plan, was approved by Cabinet in April 1998 and is now fully backed up by the Animal Breeding Act, 2001. Specific livestock sector policies include the following:

a) The National Animal Breeding Policy (1997);
b) The Animal Breeding Act (2001);
c) The Animal Feeds Policy;
d) The Meat Policy;
e) The Dairy Master Plan (1998);
f) Improvement of Delivery of Advisory Services;
g) Policy on Marketing Livestock and Livestock Products (MAAI, 2013);
h) Pasture and Rangelands Policy; and
i) Hides, Skins and Leather Development Policy.

Sudan

As reported by De Wit (2004), military conflict in Sudan has hastened a fragile and fractured set of land tenure arrangements. There is not one single legal framework recognized by all as legitimate; instead the country is divided into two largely autonomous zones. The Comprehensive Peace Agreement (CPA) of 2005 divides control of Sudan between the northern Government of National Unity (GNU) in Khartoum and the Government of Southern Sudan (GOSS) in Juba.

The legal framework of Sudan is marked by lack of clear jurisdiction between federal and state governments and legal plurality that is statutory, customary and religious legal frameworks. Land legislations at the national level favour mechanized agricultural land uses over pastoralism, while customary systems continue to operate widely in rural areas. The legislation governing natural resources is often not implemented, contradictory and of little relevance in rural areas where customary law is more highly respected. There is no legislation at the national level, which specifically regulates livestock mobility, or secures pastoralist land rights as well as rangelands (Mohammed and Faisal, 2005).

Ethiopia

According to Helland (2006), pastoralists make up 10% of Ethiopia's population, and approximately 40% of the national land area is only suitable for raising pastoral livestock. Despite this fact, there is ambiguity and contradiction towards pastoralism in the public policy discussions. Pastoralism is recognised as a viable livelihood system, but livestock mobility is considered backward and a cause of conflict and land degradation due to a poor understanding of the dynamics of dry land ecology and the rationale underpinning pastoralism.

Ideas such as livestock corridors are little understood. In the Ethiopian Constitution, only one article (out of 105) relates to pastoral areas. In addition, while the legal security of pastoral land tenure has improved in recent decades in Ethiopia, this has not been translated into more secure land tenure in practice, with alienation from valuable resources continuing. This appropriation of pastoral areas for other uses threatens the survival of pastoralism in Ethiopia and exacerbates land conflicts at large.

It is argued that some pastoral groups have lost access to sacred burial sites (Dyer, 2008). On the other hand, land tenure legislation continues to value other land uses as more productive than pastoralism, leading to widespread alienation of land. Hence, promotion of the livestock sector could have significant economic value even to achieve a livestock development programme in Tanzania, which aims to boost the contribution of the livestock sector to GDP from 3.8% in 2010 to 5% by 2016 (URT, 2011). Among other important factors to realize these objectives, political will towards pastoral systems and implementing a land policy specific to support pastoral system are necessities (Helland, 2006).

Somalia

MoPDE and MoA (2008) contend that, despite lack of international recognition, Somaliland has been functioning as a state since 1991. The elected Somaliland government maintains a relative security and peace, but instability in the region has led to illegal land grabbing and a proliferation of land conflicts. Pastoralists are the victims. In particular, they have been excluded from traditional rangelands, high value pastures have been fenced off, and livestock corridors to markets and water points blocked. The practice in Somalia is that the land rights of pastoralists are not secured in the legal system, legislation is regularly not enforced due to lack of government capacity, and there are overlapping ministerial mandates and continued influence of customary law.

An Overview of Issues for East Africa

To summarise, four governance issues stand out from the foregoing discussion. These are sectoral and contradictory legislation, legal dualism, inadequate administrative and financial capital to enforce the laws and structures of decentralization, as discussed below.

Sectoral and contradictory legislation

Hesse and Thébaud (2006) argue that almost all of the countries demonstrate sectoral and contradictory legislation in rural areas, or a lack of shared thinking in utilization of rangelands. A variety of pieces of legislation have been introduced to govern access to natural resources but each focuses on a particular sector such as forests, water, mineral resources and agriculture, that is to say legislations are disaggregated. Basically, this results in overlaps and contradictory provisions.

In most cases, the situation has marginalized pastoralist rights to access land and water resources for their activities. The multiplicity of legislation governing access to natural resources in many countries also makes it more difficult for pastoralists to understand and comply accordingly with policy governing livestock movement or mobility.

Legal dualism

Legal dualism is the situation whereby two incompatible legal codes operating side-by-side is widespread across the region. Legal dualism can be perceived as a gap between customary regulations and state legislation. This situation is further complicated in some countries in East Africa such as Sudan influenced by religious law, leading to legal pluralism. Generally, customary systems are, in the main, decaying and can no longer adequately govern new forces and demands driving land accumulation.

Likewise, the relationship between customary law and pastoralists differs significantly between regions. In areas where pastoralists have been resident for many centuries, the customary laws incorporate pastoralists and pastoral land use. This is evident in Sudan and Somaliland, but in other areas where pastoralists have a shorter history, customary governance is mainly based on settled agricultural systems.

Government capacity and implementation

Most of the governments in East Africa have inadequate administrative and financial capital to enforce legislation. This constraint is even more acute for local governments. At the national level, livestock ministries are often politically marginalized and underfunded as a result of least priority.

Countries like Somalia, for example, have enacted legislation which prohibits the enclosure or fencing of pastoral areas. However, the government cannot enforce the rules and the practice is widespread. Another factor reported slowing enforcement of the law and implementation of policies has been the gap of several years between the passing of a guideline law giving pastoralists certain rights, and the detailed regulations laying out the modalities of utilization.

Decentralization

Decentralization varies from country to country. In some there is no useful decentralization despite legislation to that effect (SWAC, 2006). In these cases, the crucial issues revolve around the level at which decentralized structures in different local authorities, the powers given to them by central government and their relationship to traditional governance structures. Generally the prevailing situation is that an appreciation of the double-sided nature of decentralization for pastoralists is becoming evident and practiced.

Conclusion and Recommendations

Pastoralism in Uganda may not be as big as it is in Kenya and Tanzania, but it is still very important in the northeast and other dry areas. However, pastoral affairs have not been accorded prominence in the government structure, as is the case in Kenya, Tanzania and Sudan. Uganda adopted a decentralized system of government under which the interests of the pastoral people can potentially be taken care of. The main problem, however, is the interest and capacity of pastoralists to participate, given the low level of knowledge/skills and remoteness from the centre of power. There are critical challenges facing pastoralists that need appropriate structures and systems to address them, especially in the very dry areas where pastoralism represents the most viable and sustainable source of livelihood and natural resource use.

Noteworthy is the fact that traditional production systems in Uganda are likely to remain important in the future. This is because the main population will be rural based and their ability to use improved breeds, in the given environment and production system, will still be constrained and hence local adapted breeds will still be kept. In addition the north-eastern part of the country may not change due to the culture, adverse climate, and land tenure system, hence there will be continued traditional production of cattle, goats and sheep.

At the East African level, pastoralism is an important economic production system and source of livelihood for arid and semi-arid areas. It is most important in Kenya and Sudan, less so in Uganda and Tanzania, mainly due to differences in the way governments are recognizing the appropriateness and importance of pastoralism at different degrees. In Kenya, for example, the constitution and laws accommodate pastoralism, including communal land tenure and mobility. Tanzania, on the other hand, recognizes pastoralism but does not support mobility. Uganda recognizes pastoralism but promotes individualization of communal grazing land.

Hesse (2000) indicates that there is great diversity in the national legislative systems and pastoral societies in East Africa. Policy and legal frameworks in the East African region are largely based on those of the former colonial power. While the French legal code is in general a more top-down model that attempts to define laws for every case, the British legal code, on the other hand, relies on more flexible, common law. Since many of the states became independent, they have experienced diverse social and political changes, conflict and development; there is also much diversity in the practices related to the handling of pastoral and agro-pastoral communities across the region analysed in this chapter.

For example, countries like Kenya and Sudan are coming up with strong policies to integrate rangelands and pastoralism into the mainstream economy. This is reflected in the Vision 2030 which includes the unique needs of arid and semi-arid lands and the establishment of a Ministry of State for the Development of Northern Kenya and Other Arid Lands. Policies in Tanzania, on the other hand, appreciate pastoralism in some policy documents and express contradictory statements in official statements and other policy documents (PASDEP, 2006).

Taking a critical look at the economic policies for Kenya, Sudan, Ethiopia and Tanzania, it can be seen that in terms of economic development planning, pastoralists have started to be recognized, though the need for rangeland is yet to be captured accordingly. The ambiguity and mixed nature of policy and policy statements, however, show that not all key actors in the policy arena have been reached; still more should be done to convince all policy makers at all levels that pastoralism is sustainable and economically viable.

Arguably, it should be noted that there is an increasing appreciation and recognition of pastoralism in the region. For instance, the African Union (AU) Policy Framework on Pastoralism and AU Framework and Guidelines for Land Policy in Africa (2010) are some of the central policy instruments providing a generic framework for countries to use to address the pastoralist issues in the region. In East Africa, the East Africa Protocol on Environment and Natural Resources, the East Africa Climate Change Policy and the Intergovernmental Authority for Development, as well as Livestock Policy Initiative are some of the key instruments providing positive guidelines on pastoral issues in the area.

Pastoralism should be unilaterally accepted as one of the most viable and sustainable uses of arid and semi-arid areas in the region and should be provided for in the laws and regulations as in Kenya. In more humid areas, more intensive agriculture, including intensive agropastoralist production, is more appropriate. However, ASALs receive low but unreliable rainfall in terms of amount, timing and location, resulting in variation in availability of the amount and location of pasture and water over the seasons and years. As shown by some of the chapters in this book, pastoralism has evolved strategies for utilizing rangelands opportunistically but sustainably for supporting livelihood. These strategies include communal land tenure, mobility of livestock, and herd management practices and rules that maximize livelihood support and food security.

Furthermore, governments should refrain from developing policies and laws or regulations that go counter to sustainable pastoralism, including individualization of grazing lands, fixed grazing and inappropriate water distribution systems. In addition, land-use planning in pastoral areas should provide for seasonal grazing patterns and, in particular, dry season grazing areas should be protected from being allocated to other uses. Finally, policies and regulations should be harmonized to regularize cross-border movement of livestock, especially during severe droughts as has been done in West Africa.

The grazing resources in pastoral areas are increasingly getting scarce for supporting a descent livelihood. Use of improved technologies may increase carrying capacity but this is of limited possibility. However, there are possibilities for diversification into non-pastoral activities within the pastoral areas and employment outside the system. Thus, displaced pastoralists for various reasons should be assisted to find alternative sources of livelihoods, build capacity of pastoralists, especially the youth, for employment in activities outside pastoralism within the pastoral areas (*e.g.* trading, tourism, rural processing, *etc.*).

Pastoral areas have received less investment in relation to need and contribution to national economies and have not been adequately consulted in national affairs, even on matters that affect them directly (*e.g.* alienating pastoral areas to conservation and privatization). This has resulted from the low level of education prevailing, remoteness from centres of power, as well as the general nonchalance of pastoralists with regard to affairs outside pastoralism. Kenya has built a more powerful and effective political lobby group by members of parliament, senators and councillors from pastoral areas.

It is recommended that representatives from pastoral communities should form effective lobby groups and should be empowered to influence policy, selection of projects and support for community initiatives. Conflicts over grazing and other resources have been intensifying within the pastoral communities and outside them in all countries of the region, resulting in death of people and livestock. The conflicts are intensified by persistent disregard to the needs and aspirations of pastoralists by the state and other institutions. There is need, therefore, for national or regional dialogues on pastoralist needs and concerns; development of appropriate mechanisms for managing conflicts between pastoralists and between pastoralists and other communities and institutions as suggested by Dr Onyango in this book; and to continue supporting socio-economic studies geared to addressing challenges facing pastoralism and developing appropriate innovations therein.

References

Behnke, R.H. (2010). The contribution of livestock to GDP in the IGAD Member States. IGAD-LPI Working Paper No. 02-10. Addis Ababa, IGAD.

Belshaw D.G.R. (1982). "An evaluation of foreign planning assistance to Tanzania's decentralization regional planning program, 1972-81." *Applied Geography*, 2 (4): 291-302

Boesen, J., Kikula, I.S. and Maganga, F.P. (1999). *Sustainable agriculture in semi-arid Tanzania*. Dar es Salaam: Dar es Salaam University Press.

Boone, R.B., J.J. Taylor, Swift, D.M. Evangelista, P.H. and Hollowed, E. (2011). Developing a resource management and monitoring protocol for a semi-arid landscape with extensive oil and gas development potential. Technical Note 439. U.S. Department of the Interior. Bureau of Land Management, National Operations Center, Denver, Colorado. 52 pp.

CBD/UNEP/IUCN (2007). Biodiversity and climate change. Montreal.

Dyer, N. (2008). Securing pastoralism in East and West Africa: Protecting and promoting livestock mobility: Review of the legislative and institutional environment governing livestock mobility in East and West Africa.

Ethiopia Ministry of Agriculture and Rural Development (MoARD), (2006). Animal Health Department. Foot and Mouth Disease Control Plan, Addis Ababa.

Fimbo, G.M. (1992). Essays in land law of Tanzania. Dar es Salaam: Dar es Salaam University Press, 138pp.

Galvin, K.A., D.L. Coppock, and Leslie, P.W. (1994). Diet, nutrition, and the pastoral strategy. Pages 113-132 in *African Pastoralist Systems: An Integrated Approach* (Eds. E. Fratkin, K.A. Galvin, and E.A. Roth). Lynne Rienner Press, Boulder, Colorado. 247 pp.

Galvin, K.A. (1992). Integrated modeling and its potential for resolving conflicts between conservation and people in the rangelands of East Africa.

GoK (Government of Kenya) (2010), The Constitution of Kenya. Revised Edition. Published by the National Council for Law Reporting with the Authority of the Attorney General- Available at https://www.kenyaembassy.com.

GoK (2007). "The fragmentation of arid and semi-arid lands" ICRC pg. 31 available at www.fao.org/.../BLBL_3Kenya.pdf.

Goodin and Northington, (1985) Plant resources of arid and semiarid lands: a global perspective.

GoU (Government of Uganda) (2001). The Animal Breeding Act, Uganda Gazette 94, Acts Supplement 36.

Hesse, C. and Cavanna, S. (2010). Modern and mobile: The future of livestock production in Africa's dry lands. London: International Institute for Environment & Development (IIED) and SOS Sahel International UK.

Homewood, K.M. (1993). Livestock economy and ecology in El Kala, Algeria: Evaluating ecological and economic costs and benefits in pastoralist systems. Pastoral Development Network Paper, No. 35a. London, UK: Overseas Development Institute.

ILRI (2006). Pastoralist and Poverty Reduction in East Africa's Conference, June 2006. International Livestock Research Institute (ILRI), Nairobi, Kenya.

Integrated Regional Information Network of the United Nations (IRIN) (2007). Africa: Can pastoralism survive in the 21st century? UN Integrated Regional Information Networks, World Press (available at http://www.worldpress.org/Africa/2861.cfm).

Kituyi, M. and Kipuri, N. (1991). "Changing Pastoral Land Tenure and Resource Management in Eastern Africa: A Research Agenda." Paper presented at the Workshop on 'Land Tenure and Resource Management among Pastoralists.' Nairobi, Kenya, June.

Land Resources Development Centre (LRDC). 1987. Tanzania: Profile of Agricultural Productivity. London: Overseas Development Institute (ODI)

Langat, R.K. (1986). "Commercial ranches in Kenya." In: Range development and research in Kenya: Proceedings of a Conference held at Winrock International Institute for Agricultural Development, edited by R.M. Hansen, R.D. Child and B.M. Wole. Morrilton, Arizona: Winrock International Institute for Agricultural Development.

Mattee, A. and Shem, M.N., Tanzania, 2005, "Consultancy report on policies and laws that affect pastoralism in Tanzania." Royal Danish Embassy Dar es Salaam and Ministry of Natural Resources and Tourism, Tanzanian Government under its ERETO 11 Ngorongoro Pastoralist Project 2nd phase.

McCarthy, N., Swallow, B., Kirk, M. and Hazell, P. (eds). (2000). Property rights, risk, and livestock development in Africa Washington, D.C.: International Food Policy Research Institute (IFPRI) and International Livestock Research Institute (ILRI). 433 pp.

Millennium Ecosystem Assessment, (2005). Synthesis Report

Ministry of Agriculture, Animal Industry and Fisheries (2013). The National Agriculture Policy-Uganda.

Ministry of Finance, Planning and Economic Development (MFP and ED) (2000). Plan for the modernization of agriculture (PMA). Eradicating poverty in Uganda. Government Strategy and Operation Framework.

MLHUD (1996). Ministry of Land, Housing and Urban Development Land and Resource tenure issues: The New wildlife Policy in Tanzania.

MoPDE and MoA (2008). Somaliland land tenure policy (2nd Draft) (LTL). Mogadishu: Ministry of Pastoral Development and Environment, and Ministry of Agriculture.

Morton, J., Livingstone, J.K., and Mussa, M. (2007). Legislators and livestock: Pastoralist Parliamentary Groups in Ethiopia, Kenya and Uganda. International Institute for Environment and Development Gatekeeper Series 131.

Mugerwa, W. (2001). Rangelands Management Policy in Uganda. A Paper Prepared for the International Conference on Policy and Institutional Options for the Management of Rangelands in dry Areas May 7 - 11, 2001 (Hammamet, Tunisia).

Mwaikusa, J.T. (1997). "The policy paper and its implications on pastoral lands." *Change*, 5 (First Quarter).

Mwalyosi, R.B.B. (1992). Land-use changes and resource degradation in Southwest Maasailand. *Environmental Conservation*, 19(2) 145-152.

Mwebaze, S. (1999). Pasture/Forage Resource Profile FAO Report available at: http://www.fao.org/ag/AGP/AGPC/doc/Counprof/PDF%20files/Uganda_English.pdf

National Bureau of Statistics (NBS) (2012). United Republic of Tanzania, Ministry of Finance January 2013. Available at http://nbs.go.tz/nbs/takwimu/references/Tanzania_in_figures2012.pd.

Neely, C., Bunning, S. and Wilkes, A. (2009). Review of evidence on dry land pastoral systems and climate change: Implications and opportunities for mitigation and adaptation. Retrieved from http://www.fao.org/uploads/media/LWdisc_paper8_temp.pdf.

Ng'ethe, J.C. (2008). Group ranch concept and practice in Kenya pastoralism: Policy legislation and governance with special emphasis on Kajiado District. University of Nairobi, Department of Range Management, Nairobi, Kenya. Available at http://www.fao.org/Wairdocs/ILRI/.

Ng'ethe, J.C. (1993). Group ranch concept and practice in Kenya pastoralism: Policy legislation and governance.

Ngana, J.O. (1993). Rainfall characteristics and their relevance to agricultural planning in semi-arid central Tanzania. In: Proceedings of Planning Workshop SWMP, 13- 15th January 1993. SU Morogoro Tanzania, edited by N. Hatibu and T. Simalenga. Morogoro: SWMP.

Njombe, F. and Msanga, N. (2008). "Livelihoods and rural poverty in Tanzania." *World Development,* 31 (8) 1367-1384.

Nyerere, J.K. (1962). "The Second Scramble 1962" Nyerere's Speeches. Available at: http://www.juliusnyerere.org/resources/view/the_second_scrumble_1962

Ococh, A.E. (2013). Uganda country paper on animal genetic resources: A presentation to the Inception Workshop of the Animal Genetic Resources (Angr) Project in Strengthening the Capacity of African Countries for Conservation and Sustainable Utilisation of African Animal Genetic Resources. Abidjan, Ivory Coast.

OP (2011). National Policy for the Sustainable Development of Northern Kenya and other arid lands. Unlocking our full potential for realization of the Kenya Vision 2030. Nairobi: Ministry of Devolution and Planning; available at http://www.fao.org/.

Oxfam (2008). Survival of the fittest: Pastoralism and climate change in East Africa. Oxfam Briefing Paper 116.

Oxfam (2006). Delivering the Agenda: Addressing chronic underdevelopment in Kenya's arid lands. London, Oxfam.

Pitblado J.R., (1970). A review of agricultural land use and land tenure in Tanzania. Bureau of Resource Assessment and Land Use Planning, University of Dar es Salaam.

Regional Pastoral Livelihoods Advocacy Project (2009b). Pastoralism, policies and practice in the Horn and East Africa. Coalition for European Lobbies on Eastern African Pastoralism, April.

Scoones, I. (2007). Climate change and the challenges of non-equilibrium thinking – Contribution to WISP e-Conference, February.

Sendalo, D.S.C.A. (2009). Review of land tenure policy implications on pastoralism in Tanzania. Ministry of Livestock Development and Fisheries, Department of Livestock Research, Training and Extension. Available at https://www.unpei.org/ accessed May 2016.

Shechambo, F.C., Sosovele, H. and Kisanga, D. (1999). Rethinking natural resource degradation in semi-arid Sub-Saharan Africa: The case of semi-arid Tanzania. Institute of Research Assessment, University of Dar es Salaam, Tanzania. Sourced at http://www.odi.org.uk/RPEG/soil_degradation/tzlit.pdf.

Shem, M.N. (2010). Impacts of government policies on pastoralist livelihoods in the semi-arid areas of Tanzania. Paper presented atICID+18, August 16 - 20, 2010, Fortaleza- Ceará, Brazi on line: http://www.dewpoint.org.uk accessed February 2016.

Shivji, I.G. (2004). "Maasai rights in Ngorongoro, Tanzania." IIED/ HAKIARDHI Tanzania. Environmental Conservation, 19(2): 145–152.

Tenga, R. et al. (2008). Legitimizing dispossession: The Tanzanian High Court's decision on the eviction of Maasai pastoralists from Mkomazi Game Reserve. Lands, Resources and the environment.

Tenga, R. (1996). Land tenure in Tanzania. In: Leader-Williams N., Kayera J.A. and Overton G.L. (eds.) Community-based conservation in Tanzania. Proceedings of a Workshop Held in February 1994. IUCN, Gland, Cambridge, pp. 19–28.

Thornton, P.K., Jones, P.G., Owiyo, T., Kurska, R., Herrero, M., Orindi, V., Bhadwal, S., Kristjanson, P., Notenbaert, A., Bekele, N. and Omolo, A. (2006). "Climate change and poverty in Africa: Mapping hotspots of vulnerability." African Journal of Agricultural Economics, 2 (1).

UNDP (2006). "Making markets work for the poor" (available at http:// www.undp.org/drylands/docs/marketaccess/Making_Markest_ Work_for_Poor.pdf)

UNIDO (2012). Tanzania red meat value chain: A diagnostic. Africa Agribusiness and Agroindustry Development Initiative (3ADI) Reports. UN Industrial Development Organization (UNIDO), Vienna, Austria.

URT (2012). Tanzania Livestock Policy. Dar es Salaam: Ministry of Livestock and Fisheries.

URT (2006). National Livestock Policy. Dar es Salaam: Ministry of Livestock and Fisheries Development.

URT (2005). Livestock Sector Development Programme. Dar es Salaam: Ministry of Livestock and Fisheries Development.

URT (2004). Aligning and harmonizing the livestock and land policies of Tanzania. Dar es Salaam: Ministry of Livestock and Fisheries Development.

URT (2001). Poverty Reduction Strategy Paper (PRSP). Dar Es Salaam: Government Printer.

URT (1999). The Local Government Laws Act, 1998. Dar es Salaam: Government Printer.

URT (1998). The Constitution of the United Republic of Tanzania of 1977. Dar es Salaam: Government Printer; 142pp.

URT (1994). Report on existing legislation pertaining to the environment. Division of Environment, Dar es Salaam.

Veit, Peter (2011). The rise and fall of group ranches in Kenya. Regional Pastoral Livelihoods Advocacy Project, "Pastoralism, policies and practice in the Horn and East Africa," April 2009.

Watkiss, P., Downing, T., Dyszynski, J., Pye, S. *et al.* (2011). *The economics of climate change in the United Republic of Tanzania.* Report to Development Partners Group and the UK Department for International Development. Available at: http://economics-of-cc-in-tanzania.org/.

Watkiss, P., Downing, T., Dyszynski, J., Butterfield, R., Devisscher, T., Droogers, P., Pye, S., Ali, B., Harding, B., Tas, A., de Blois, M., Tadege, A., Hunt, A., Taylor, T., Bouma, M. Kovats, S., Maitima, J., Mugatha, S., Kariuki, P., Mariene, L., Worden, J., Western, D., Waruingi, L., Brown, S., Kebede, A., Nicholls, R., Lager, B., Otiende, B., Chambwera, M., Birch, T., Mutimba, S., Sang, S. (2009). *The economics of climate change in Kenya.* Final Report to DFID and DANIDA. Led by the Stockholm Environment Institute, Oxford. Available at http://kenya.cceconomics.org/kedo/FINAL-kenyareport-April.pdf.

World Bank (2007). Temas de conversacion. Washington, DC: World Bank. http://documents.worldbank.org/curated/en/492411468134394897/Temas-de-conversacion-April-de-2007

World Bank (1992). Tanzania: An historical review. Pastoral Land Tenure Series, No. 8, IIED, London.

PART IV

Analysis of the Contribution of Pastoralism to the National Economies of East Africa

CHAPTER 10

Pastoralism in East Africa: Economic Trends, Patterns, Opportunities and Challenges

Edmund Mabhuye

Centre for Climate Change Studies, University of Dar es Salaam, Tanzania

Background

Pastoralism as a production system has been conceptualized differently in various literatures. In the 1940s, for example, Evans-Pritchard (1940) defined the pastoral production system by emphasizing herding of domesticated animals as a major subsistence activity. Later day definitions are, however, more quantitative, preferring Swift (1988)'s definition that "...the pastoral production system is a type of production system in which at least 50% of the households' value of market production and the estimated value of subsistence production consumed by households result from herding of domesticated livestock pastoralism or its related activities, especially production of meat, milk or dairy products". However, this definition does not capture the differences that exist in various countries where pastoralism is practiced and perceived differently depending on their political, cultural and economic settings.

According to Hatfield and Davies (2006), pastoralism may not always fulfil all criteria, since some pastoralists may shift, from year to year, between varying degrees of cultivation or off-farm labour that might periodically define them as multi-activity pastoralists. In general, however, a majority of the definitions limit themselves to some aspects of pastoralism, such as nomadism, regardless of the extent to which it contributes to the household economy. Hence, to these definitions pastoralism refers to any predominantly livestock-based production system that is mainly extensive in nature and uses some form of mobility of livestock (Hatfield and Davies, 2006).

As clearly shown by Gaiballah and Abdalla, and others in this book, East African pastoralism is a livelihood and production system that is practiced in the ASALs. It is a form of production that manages a complex network of profitable cross-border trade and draws huge economic benefits from rangelands ill-suited to other land-use systems (IIED and SOS Sahel UK 2010).

Pastoralists explore unproductive lands which cannot sustain other forms of production, partly because of inherent environmental constraints, mainly extreme climate variability. Pastoralists, however, use climatic variability to their advantage to maximize the productivity of their herds. As shown in earlier chapters of this book, they adopt various forms of strategies to enable them to make economic use of these lands without doing harm to the environment. The strategies include livestock mobility and selective breeding of animals capable of tracking highly dispersed concentrations of nutrients in the rangelands (Benjaminsen *et al.*, 2009; IIED and SOS SAHEL, 2010).

Some authors have named pastoralism as the huge hidden asset, which has not been well tapped in the stimulation of national and local economic development (IGAD, 2010). Although the structure of the economy and the framing of national economic policies varies considerably between countries, these 'hidden assets' are acknowledged by some authors to have immense potential value for reducing poverty, generating economic growth, managing the environment, promoting sustainable development and building climate resilience (IIED and SOS Sahel UK, 2010).

Despite the immense economic potentials of the pastoral production system, many policy makers in East African countries have a blind spot regarding pastoralism, in particular its contribution to economic growth. Existing data significantly undervalues the total economic value of pastoralism to national and regional economies (IIED and SOS SAHEL, 2010; Benjaminsen *et al.,* 2009). Available statistics for nationally-produced goods and services only reflect pastoralism's *direct economic values* (products like milk, meat, livestock, hides, leather products, directly sold or sold as inputs to national and international supply chains). However, behind such direct economic values there are also the *indirect economic values* of the herd, such as the social and cultural values of livestock that have largely been ignored. Therefore, analysis of the pastoral economies in East Africa is imperative in order to highlight these indirect values and inform policy, and ultimately better understand and enhance pastoralism to yield significant contributions to local and national economies in the East African region.

Objectives of the Study

The main purpose of this study was to collect data that would enable us to provide a critical analysis of the status of knowledge in the pastoralist economies by linking the global, regional and local issues relevant for the thematic areas addressed by this chapter. The analysis focuses on the contribution of pastoralism to national economies. It documents the nature and patterns of pastoral economies, and partially analyses the policy and institutional framework governing pastoral economies, and how it influences their contribution to the local and national economies of the East African region.

Methods and Approaches

Development of this book chapter relied mainly on the review of literature with a mix of stakeholder consultations to validate findings from the literature review. Consultations with stakeholders in selected districts in Manyara and Arusha regions aimed at collecting additional information for validating analysis from the literature review. Stakeholder consultations employed various methods, including key informant interviews and focus group discussions. Focused field surveys were conducted in Monduli, Simanjiro, Longido and Ngorongoro Districts.

Literature review used a systematic approach in order to adhere to the current methods recommended for the reviews. By definition as perceived in this review, systematic reviews are summaries of research evidence that address objectively formulated questions using logical and clear methods to identify and critically appraise relevant research methods (Cooper, 1998). Systematic reviews also collect and analyse data from studies that are included in the review as an important role of providing evidence used to inform different stakeholders, including policy makers. The reviewed literature in this case included peer reviewed journal articles, workshop proceedings, books, published and unpublished reports.

Materials for review were obtained from electronic and physical databases. The search enabled us to obtain published and unpublished research reports with relevant information as per the objectives of the review. The review also included cross-checking reference lists of all full text articles included in the review and searched accessible grey literature. The following subsections explain specific approaches used during the literature review.

Inclusion and exclusion criteria

All quantitative and qualitative studies relevant for the analysis of the contribution of pastoralism to the national economies and wellbeing of pastoralists in East African countries were analysed. For the sake of this review, pastoralism was defined as an economic, social and cultural activity that has existed for centuries in different communities globally. In order to be included in the literature under review, the studies had to focus on issues related to how pastoralism can be enhanced for better livelihood and national development in East Africa, as well as how framing structures such as policies, legal and institutional framework enhance and/or constrain the performance of pastoralism in the national economy and for the wellbeing of pastoralists in the East African region and in Tanzania, in particular.

Quality assessment

The risk of bias was appreciated in order to assess the quality of studies included. Each study was evaluated, guided by the criteria set for inclusion and exclusion. In case of any discrepancy in the perspectives of a study, a final decision was reached guided or informed by the objectives set for the development of the book chapter. In summary, all the studies included in this review were of strong quality, with low to moderate risk.

Data extraction

Extraction of data from each article included in the study was done by using a data collection form. Information extracted from each study included the citation, study design and methodology, geographic setting, economics of pastoralism in relation to the national and local economies, patterns and trends, as well as the wellbeing of pastoralists, outcomes and findings. Also, extraction of information was guided by the objectives of the book chapter and inclusion or exclusion criteria.

Dealing with missing data

In cases of missing or incomplete information presented in the reviewed literature, the author attempted to fill the gaps by collating additional information through stakeholder consultations and field visits. Narrative synthesis was done of the relevant data culled from each of the methods mentioned above and used in writing this chapter. The main findings of the study are presented in the subsequent sections.

Findings of the Study Patterns and Trends

Varied interpretations and perceptions

Various writers acknowledge the existence of pastoralism and its social and economic contribution to every sphere of human life. Ndagala (1990) identified pastoralism as an economic, social and cultural activity that has existed in East Africa for centuries – many years even before the advent of colonialism. In fact, Ndagala (1990) notes that writings of early agents of colonialism like Krapf and Carl Peters also disclose the fierce and hot confrontations they encountered when they were passing through pastoral communities in various parts of pre-colonial East Africa.

As classified by Khazanov (1984), the economic characteristics of pastoral societies are varied based on animal species, management system, geographical location and political ecology of an area. Similarly, Trabucco and Zomer (2009) indicated that the rangelands in East Africa are not homogencous and their climates range from hyper-aridity to sub-humidity, specifically varying considerably in rainfall variability, soil characteristics and soil fertility.

In this context, one study shows that variations in terms of economic, social and demographic characteristics (whereby population densities in these rangelands may range from almost zero to more than 200 people per sq. km. are a normal characteristics in pastoral communities (CIESIN *et al.,* 2004). Rangeland landscapes and the communities surrounding them are not all the same (Blench, 1998), so they will respond differently to the impact of climate change. This complexity and heterogeneity of socio-economic characteristics needs to be closely analysed in the context of climate change.

According to Benjaminsen *et al.* (2009), pastoralists in many ways are considered as people who are aggressive, warlike, rigid, autonomous and economically superior to their neighbours. Another perception is that some pastoralists, especially the agro-pastoralists, though economically superior are cooperative with their neighbours. These perceptions are not only held by the neighbouring communities but also by the local government and the state (*ibid.*).

At another level, Dyer (2008) noted that government in most countries perceives pastoralism and pastoralists as an archaic institution that knows nothing about modern ways of keeping their herds; thus they need to be taught better and modern ways of raising their livestock.

Consequently, partly due to this perception, as reported by various researchers, there have been top-down planning and implementation of livestock related programmes in various countries of East Africa (Dyer, 2008; Benjaminsen *et al.*, 2009). It is thus that there have been a number of cases reported on the confiscation and conversion of land and other resources they have used for generations. This is partially made vivid through a number of conflicts between pastoralists and other land users, which are alarming in various parts of East Africa (*ibid.*).

Pastoralists as contributors to economic development

The recognition of pastoralists and the contribution of livestock to economic development and improvement of livelihood to pastoralists are among the issues that have attracted the attention of different researchers and stakeholders (Melville, 1994). Major concern has been pressurizing governments and other stakeholders to join the effort to recognize the role of pastoralism in economic development. The world's livestock population was for a long time perceived as an underutilized resource and thus stimulated research and formulation of various development projects.

Analysis of the contribution of pastoralism in the national and local economy also influenced undertaking this study. Another major concern of researchers and other key players for the pastoralism sub-sector has been the impacts of climate change which have become a serious threat to marginalized and vulnerable pastoralists worldwide. Key players are increasingly becoming alert about the sustainable management of arable land, grasslands and rangelands which deserve greater attention now for sustainable pastoral and agro-pastoral livelihoods for millions of people in the world (FAO, 2009).

In this context, Hatfield and Davies (2006) indicate that pastoralism has been evolving and shaped by many distinctive factors, including climate change impacts, which have surpassed the capacity of pastoral communities to withstand challenges associated with climate change. Drylands degradation, which not only erodes natural resource endowment for pastoralism, but also escalates poverty among pastoral communities, has aggravated the severity of climate change impacts, driving to loss of pastureland and livestock endowment, especially during the dry season. These factors are among many other reasons for declining trends of pastoralism and economic endowment and increasing vulnerability to climate change impacts; they have been widely documented, and have attracted the interest of many researchers. As reported by Hatfield and Davies (2006), marginalization of pastoralism also hampers efforts such as policy interventions to return pastoralism to its prior or imagined state.

Although some researchers like Hatfield and Davies (2006) recommend mobile pastoralism as a sustainable solution to the problems of resource degradation and climate change impacts in drylands, given the greatest economic potential which lies in mobile pastoralism, however, mobility of pastoralists especially in most parts of Tanzania have resulted in conflicting land uses and clashes among farmers and pastoralists (Ojalammi, 2006; Dyer, 2008; Benjaminsen *et al.,* 2009). The situation is explained as linked to lack of appropriate policies, legal mechanisms and support systems (Hatfield and Davies, 2006) in order to enhance the mobility and support economic, social and ecological sustainability of their livelihoods.

Given the challenges, many pastoralist communities are now increasingly sedenterizing (for service and market access) (Hatfield and Davies, 2006), sometimes devoting labour to small-scale cultivation in response to market dynamics and policy influence as a way to increase their productivity. Again, this option needs to be promoted and supported through policy interventions to improve infrastructure for fattening their livestock in order to have a profound impact on livestock marketing (Hatfield and Davies, 2006).

Nevertheless, producing milk for consumption remains a central part of the livelihood of many pastoralists, raising challenges for sedentary communities and commoditization of the pastoral economy (Muhereza and Ossiya, 2004; Hatfield and Davies, 2006). Noticeable change in the pastoral economy has been diversification of livelihoods partly to reduce exposure to natural resource fluctuations, and engagement in cash generating activities, driven by a combination of changing aspirations in education, access to information and modern consumer goods. In many cases, off-farm and non-farm income earning is being adopted in lieu of increased time investment into livestock production and marketing (Desta and Coppock, 2005). Value addition to pastoral products is increasingly becoming common (*e.g.* butter and other livestock related commodity manufacture).

Livestock and wildlife interaction

Despite the human-wildlife conflicts that threaten co-existence of wildlife and livestock (Case Study 9.1), conservationists are increasingly exploring ways to work with pastoralists for the conservation of the extensive rangelands. The recent change in understanding of range ecology has convinced many people that pastoralism is necessary for effective rangeland management, and to continue with policies that undermine pastoralism or with attitudes that alienate pastoralists is to sow the seeds of failure of conservation (Boyd, *et al.* 1999; Nelson and Ole Makko, 2005). Several examples can be found in East Africa of pastoralists grazing their stock in national parks and benefiting from tourist revenues (Hatfield and Davies, 2006; Homewood *et al.*, 2012).

Case Study 10.1: Simanjiro District

In Terrat Village, Simanjiro District, wildebeest move out of Tarangire National Park and go into the open plains within the village lands during the month of October where calving and breeding takes place. The wildlife roam the open plains and go back to the National Park in June the following year. The October migration coincides with the rainfall period when pastoralists are creating pasture reserves for use during the dry season. However, wildebeest graze the entire area and also break into enclosures. This reduces pasture availability, leading to early migration of livestock to other villages even during times of minimal drought. Wildebeest during the calving period spread malignant catarrhal fever, a disease that kills a huge number of livestock. Pastoralists thus avoid the open ranges occupied by wildlife as a result of disease prevalence.

The biggest problem in this situation is that pastoralists on the contrary cannot access pasture for their livestock in the National Park during times of need due to existing policy and legislation bottlenecks that prohibit access to these areas which are classified as biodiversity conservation areas, unlike in the neighbouring country of Kenya. This often leads to human-wildlife conflicts (**Source:** PINGOs Forum, 2016).

Wildlife Management Areas (WMAs) such as Makame and Burunge, which have been established in Tanzania to allow co-existence of wildlife and livestock reveal successes, despite some few challenges related to management strategies (Sulle *et al.* 2011; Tetra Tech and Maliasili Initiatives, 2013). Generally, pastoralism in East African drylands is undergoing rapid transformation (Greiner *et al.*, 2013). Despite the promising trends, however, it is still heavily dependent on unprocessed primary products, and thus its impact to the economy is not optimal (Wood and Mayer, 2001). This calls for immediate interventions to promote processing of livestock products in order to improve pastoral earnings and diversification of their economic systems.

The need for diversification of pastoral economic systems, especially in the context of climate change, arises from the fact that despite inappropriate documentation, pastoral systems have a major role to play in economic development and, importantly, in supporting pastoral livelihoods. For example, UNDP (2006) shows that livestock keeping is of great significance, both economically and socially, to rural livelihoods, thus high priority should be given to facilitate sustainable pastoralism. UNDP (*ibid.*) has gone much further by elucidating that pastoralism provides for livelihoods, ensures ecosystem services, and promotes wildlife conservation. It also honours traditions and cultural values in pastoral communities.

Despite this growing trend of analysis of the practice of pastoralism, it is obvious that climate change will negatively impact the ecological resilience of pastoral rangelands and thus require immediate responsive measures to implement adaptation activities (The economic impacts of these changes in East Africa, including Tanzania, are well discussed in Watkiss *et al.*, 2009; 2011). Nevertheless, success of adaptation efforts will depend on recognition of the importance of livestock within total household income, and in the local and national economy.

Recognition of the value of pastoralism to economic development can also be motivated by appropriate documentation of the value of pastoralism as an asset for economic development. Engagement of key stakeholders and players across scales and levels may improve the process and influence policy change in order to devise pro-poor livestock policies, which address their immediate and long-term needs, as well as natural and social dimensions for stimulating economic growth and sustainable pastoral development.

Contribution of pastoralism to local and national economies

Many authors consider pastoralism a hidden asset (IIED and SOS Sahel, 2010), made up of millions-strong herds of livestock managed by pastoralists. A significant, but unknown proportion of the national livestock herd in East Africa is raised in pastoral areas. Pastoralists manage complex webs of profitable local and cross-border trade, which as a result draw huge economic benefits to both the community surrounding them and themselves. For centuries, livestock have fed families and led to growth of the economies. However, the total economic value of this asset is not treasured (Homewood *et al.*, 2012).

Precise figures for the contribution of pastoralism to local and national economies in East Africa are relatively hard to come by. The visible contributions of pastoralism to the economy are generally limited to the measurement of livestock sales and some by-products, such as dairy milk, meat and hides (Hatfield and Davies, 2006). Whilst these are challenging to quantify in themselves, they do not capture the full value of pastoralism, since there are multiple and extensive sets of values associated with pastoralism, some of which can be measured but many cannot; and those that can be measured are often underestimated (*ibid.*).

The majority of pastoralists consume milk and produce a range of dairy products, yet marketing of those goods is often limited or restricted. Another economic value, which has not been valued, is the power for transportation and animal traction, which are provided by livestock for many pastoralists (Waters-Bayer and Lossau, 1995; Hatfield and Davies, 2006). This is partly linked to the policy decisions that affect pastoralists' information and their existing values (FAO, 2003; Hesse and MacGregor, 2006). As a result, Hesse and MacGregor (2006) suggest that economic values of pastoralism can only be estimated if the multiple values of pastoralism are understood and accounted for, whether or not they have a market value and whether they are produced or foregone.

This means that an accurate estimation of economic values of pastoralism can only be achieved through a holistic valuation of all the goods and services that pastoralism provides, by looking beyond the immediate benefits of livestock and livestock products. As observed by many authors, the Total Economic Value (TEV) framework is one of the important tools that could provide information for understanding the true contribution of pastoralism to local and national economies (Hesse and MacGregor, 2006; Hatfield and Davies, 2006).

Using the direct benefits analysis, some previous studies have indicated that in countries with a dominant agricultural sector, pastoralism can be a significant contributor to the national GDP (Hatfield and Davies, 2006). Globally, livestock is growing faster than any other agricultural subsector, and it is predicted that by 2020 it will produce about 30% of the value of global agricultural output (Delgado *et. al.*, 1999). The World Bank (2007) estimated that livestock in some developing countries accounted for about 50 to 80% of the GDP. Furthermore, FAO (2009) indicated that livestock production was a growing sector and was estimated that 1 billion people depended on it, and livestock served as a source of income and food security to about 70% of the world's 880 million rural poor, the majority of which were found in Sub-Saharan Africa.

In Uganda, for example, Hatfield and Davies (2006) reported that pastoralism was contributing around 8.5% of the total GDP, mostly being contributions from foreign exchange earnings from sale of direct benefits (Muhereza and Ossiya, 2004; IIED and SOS Sahel, 2010). To date, livestock keeping continues to be the fastest growing agricultural sub-sector in Africa and Tanzania, in particular (URT, 2006; Engida et al., 2015). For instance, though the statistics might have changed or remained constant, by 2006, Tanzania had the third largest cattle population in Africa after Ethiopia and the Sudan (URT, 2006; NBS, 2012).

In Kenya, livestock raised by pastoralists is worth USD $800 million a year (IIED and SOS Sahel, 2010). The sector is worth an estimated three-quarter of a billion dollars a year, while in Tanzania the sector in the financial year 2014/2015 alone contributed directly to the economy an amount of Tsh. 6,486,534,052.89 as revenue from the livestock sector, leave alone other economic benefits like employment and the food security livestock provided to the people (Letara, James and Hesse, 2006; URT, 2015).

Exports of livestock and livestock products from the Horn of Africa now annually exceed USD $1 billion and more than 90% of this commerce derives from production systems based on mobile pastoralism (Little, 2013). In Ethiopia, the livestock sector ranks second to coffee in generating foreign exchange. In 2006, for example, Ethiopia earned USD $121 million from livestock and livestock products (National Bank of Ethiopia database 2010; IIED and SOS Sahel, 2010).

However, apart from the envisaged significant contribution towards achieving the development goals of the National Strategy for Growth and Reduction of Poverty (NSGRP), the livestock sector has been unstable for various reasons. These have included factors such as environmental change. The sector's contribution to the GDP has been unstable as it declined from 18% to only 5.9% between 2001 and 2005 (Economic Survey, 2005; URT, 2010). The livestock sector contribution to the agricultural GDP was about 3.8% in 2010, and contributed 4.0 % of the national GDP in 2009 compared to 4.7% in 2008 (URT, 2010; 2011).

Despite these declining trends, the livestock sector has remained significant to the economy if other benefits are considered. New research shows, for example, that pastoralism has immense potential value for reducing poverty, managing the environment, promoting sustainable development and building climate resilience (IIED and SOS Sahel, 2010). The other important benefit of the livestock sector, in this respect, includes the contribution to crop farming as a source of manure and draught power. It also fulfils socio-cultural roles in livestock farming communities (URT, 2010). This has been one of the policy challenges associated with undervaluing of the contribution of livestock to the national economy (Engida *et al.*, 2015).

IRIN (2006) estimated that 40% of the land in Africa was dedicated to pastoralism and 70% of the population relied on dry and sub humid lands for their daily livelihoods. An estimated 50 million pastoralists and up to 200 million agro-pastoralists live across dry land Africa from west to east (IIED and SOS Sahel, 2010). However, these dry lands, which are predominantly used for livestock production, are particularly vulnerable to the impacts of climate change.

It is also estimated that around 23% of the world's poor (nearly 300 million people) are located in Sub-Saharan Africa. About 60% of these depend on livestock for their living (FAO, 2009). Furthermore, 25 million pastoralists and over 240 million agropastoralists depend on livestock as their main source of income (IFPRI and ILRI, 2000). An estimated 50 million livestock producers in East and West Africa support their families and communities through livestock (Behnke and Muthami, 2011). Massive meat, skins, hides and milk industries are established which depend on obtaining raw materials from the livestock kept by pastoralists in the regions.

Where other economic systems are failing due to unpredictable weather, climate change and other challenges, pastoralism has continuously generated huge national and regional economic benefits due to its sustenance and resilience strategies (URT, 2006). The majority of pastoralists (over 90% in Tanzania) depend on indigenous breeds. Despite their low genetic potential, these animals are well adapted to harsh environmental conditions and have high resistance to diseases (URT, 2006).

Pastoralism offers the most cost-effective way of supporting relatively large populations in these areas and at minimal environmental cost. Furthermore, and perhaps more significantly, it offers East African countries a major opportunity to capitalize on the rapid projected growth in demand for livestock products over the next 15-20 years. The continued neglect of pastoralism carries huge potential costs as poverty, environmental degradation and conflict are likely to increase as local people lose their livelihood base and struggle for survival (Behnke and Muthami, 2011).

A prevalent perception about pastoralists is that it is a minority of people who are practicing an outdated lifestyle. Pastoralism is also seen by many as archaic, economically inefficient, chaotic and environmentally destructive – perceptions that are not evidence-based, yet prevail and drive most of the views of leaders and regional policy (Benjaminsen *et al.*, 2009). Inadequate and inaccurate national statistics on pastoralism do little to alter this prevalent perception.

Persistent undervaluation has effectively trapped millions of pastoralists in a cycle of poverty, conflict and vulnerability to the impacts of climate change. However, the truth is that, even though pastoralists inhabit harsh and remote regions, they are fully integrated in the wider global processes. Rapid population growth with rising urban incomes has led to high demand for meat and dairy products. It is mainly pastoralists who are meeting the satisfaction of this demand. As illustrated in the subsequent sections, livestock have both direct and indirect benefits which vary significantly.

Direct benefits

Livestock products, stock accumulation, insurance and inheritance, to mention a few, are the values that are very direct from the livestock sector. The most obvious direct livelihood values are the production of milk, meat, skin, blood, *etc.* The herd is the basic unit of production which provides benefits to families, individuals and the community as a whole. Another direct benefit of livestock is its value as a household's assets or investment (Behnke and Muthami, 2011).

Investment in livestock is often an opportunity available for rural people with no access to reliable banking systems. Livestock can be sold as "stock" when the price is stable and then converted into other commodities such as grain. Like monetary stock, values can go up and down. However, most livestock appreciate over time through births, growth and maturation of animals.

It has been proven that returns of investment in livestock are often higher than bank returns and pastoralists tend to recycle their earnings by purchasing more stock. Herd splitting and other pastoralist management strategies are used to avoid stock loss through drought and disease (Behnke and Muthami, 2011).

The third direct benefit is the "insurance policy" aspect of livestock. Much benefit is placed in the herd's total size: the bigger the number of animals a family owns, the greater the chances of addressing risks and surviving uncertainties. During drought, for instance, pastoralists with larger and diverse age and species herds are more likely to recover and survive than those with small herds and one or very few species.

The fourth direct value is that livestock plays the role of a currency as a long term security. Livestock inheritance within families is critical for the establishment of new families and the survival of pastoral society as a whole. Inherited animals, which ensure security for new families, support the expansion of communities. Also, the strategy of managing grazing resources through constant negotiation of use rights contributes to building the huge social capital needed for communities to exist in high risk climate-dependent environments.

Another livestock benefit is the economic values which are associated with pastoralism and livestock keeping in general. The key benefit here are the goods and inputs marketed to other supply chains which go along with the direct subsistence value of goods and services produced through pastoral production. The substantial economic benefits of these goods in the formal market are seen through the sale and export of milk, livestock, hides and leather.

Moreover, pastoralism provides inputs to a wide range of formal and informal industries established in countries such as meat and restaurant businesses. The study by Letara *et al.* (2006) on beef related businesses in Arusha Municipality informed that it supported employment of a large number of people and this had a direct benefit to the local economy and other ancillary businesses such as abattoirs and butcheries that have significant annual turnover in the sub-sector.

Letara *et al.* (*ibid.*) indicates that by 2006, there were about 601 roasted meat (locally known as *nyama choma*) businesses in Arusha, covering a population of 800,000. The study also reported that the *nyama choma* business was estimated to provide a minimum of 5,855 full-time job opportunities, employing an estimated 2-3% of the total working age population of Arusha.

The study also reported that in 2005, the total contribution of livestock related businesses to the municipality coffers was about Tsh.177 million, which was approximately 10% of the total municipal income. Apart from contribution of the livestock sector to the national GDP, there has been considerable contribution of the sector to the local economy. Nationally, by 2006, the annual turnover of the livestock sector in Tanzania was estimated to yield about USD $22 million (Letara *et al.*, 2006; IIED and SOS Sahel, 2010).

In another study, it was indicated that cattle trekking for over 450 kilometres from southern Somalia accounted for 26% of the beef consumed in Kenya (IIED and SOS Sahel UK, 2010); 16% of that was consumed in Nairobi alone (Little and Mahmoud, 2005). The IIED and SOS Sahel UK (2010) study also indicated that the traditional livestock sector in Tanzania produced 70% of the country's milk, which was 770 million litres in 2006 (URT, 2006). Similarly, apart from unregistered livestock exports to Kenya and Ethiopia, traditionally the livestock sector has been central to the economic development of the Somali peoples, with approximately 40% of GDP, almost 90% of total agricultural GDP, and more than 60% of the population supported for food and income (Knips, 2004; World Fact book, 2011; FSNAU, 2011; Little, 2005; Maystadt and Ecker, 2014).

Elsewhere in East Africa, livestock also have had the potential in human capital benefits. The key benefits under this important category for economic development have been employment and supply of indigenous knowledge. In most cases, the direct benefit of livestock as an employer is always overlooked. In East Africa, estimates of the pastoralist population vary from 9 to 20 million and, with potential part-time involvement, the population could be considerably higher. In addition, more that 60% of the people employed in the livestock sector are adults of working age who are gainfully employed in gaining livestock and subsidiary activities such as livestock trading.

In the rural areas of the ASALs, pastoralism and agro-pastoralism are often the only form of employment. Pastoralism's input to other supply chains provides employment to urban peoples like traders, transporters, and other vendors. Pastoralists who lose their cattle face unemployment and rural-urban migration, which as a result create a very tangible cost to the national economy (De Jode, 2010).

Indirect benefits

Economic input values

According to Hesse and MacGregor (2006), agriculture is a key beneficiary of pastoralism. It helps raise agricultural productivity by providing manure, animals for agriculture and transport, seasonal labour, and technical knowledge for the rising number of farmers now investing in livestock. Farmers also help pastoralists by providing crop residues as fodder, potentially crucial in drought years. These reciprocal exchanges help reduce conflict and promote peaceful relations in agropastoral communities. Brooks (2006) and Quan and Dyer (2008) also indicate that reciprocal relationships between pastoralists and agriculturalists could play a key role in future development, including reinforcement of food security.

Tourism

A vital input from pastoralism is the maintenance of grazing reserves, which provides critical dry season habitats for wildlife. Northern tour operators in Tanzania also market trips using images of pastoralists, while pastoralist cultural performances and handicrafts have clearly helped spark interest in the region. As a form of land use, pastoralism has also helped protect the many national parks that East African tourism depends upon. Unlike agriculture, pastoralism is one of the few land uses able to coexist with wildlife, as domesticated and wild animals exploit different ecological niches.

For instance, the Maasai pastoralists are believed to have significant contribution towards the existence of the Ngorongoro Conservation Area (Hesse and MacGregor, 2006) and other national parks such as the Serengeti National Park. It is estimated that the annual value of pastoralist land management practices to the wildlife based tourism industry in northern Tanzania of approximately $85 million (Nelson and Ole Makko, 2005; Sulle et al. 2011). Revenue generated from tourism based on pastoral lands represents 50% of Kenya's GDP and over 19% of Uganda's.

Environmental Values

There are also the environmental values provided by pastoralists as custodians of dry land environments. These include nutrient recycling, maintenance of pasture productivity and biodiversity, tree regeneration, maintenance of natural ponds and carbon sequestration for both climate change mitigation and adaptation. Maintenance of grazing reserves, and other pastoralist management practices, helps to conserve wildlife on pastoralist lands.

More than 80% of Kenya's and Tanzania's protected areas are in pastoral lands. Tarangire is part of the vast Maasai Steppe ecosystem, more than 80% of which falls on lands managed by pastoralists, including key wet season dispersal habitat for wildlife such as the Simanjiro Plains (Nelson and Ole Makko, 2005). In the Serengeti ecosystem, state protected areas cover more than 25,000 km2, and the migratory herds of wildlife spend only about 3% of their time on pastoral lands.

Grasslands store approximately 34% of global terrestrial stocks of CO_2 – a service worth $7 per hectare (McGahey et al., 2014). African grasslands extend to 13 million square kilometres and have vast carbon sequestration potential. But some 18-28 billion tonnes of carbon may have been released from dryland soils through degradation resulting from non-pastoral activities. All of these go unrecognized in the national accounts.

Case Study 10.2: Simanjiro District

Simanjiro District has a large livestock population, which is largely unexploited (SDC, 2016). The district is endowed with the largest number of cattle (around 347,489 cattle as of February 2016) as compared to neighbouring Kiteto, Babati and Hanang, and Mbulu Districts in the Manyara Region. Other major types of livestock in the district include goats, sheep, and donkeys. Interviews with the district livestock officers indicated that if these livestock were being exploited, then the district could have been earning significant contribution to the development of the district through investments in small and medium scale industries for processing of livestock products such as leather, milk and milk products and meat. From the interviews it was also reported that the number of livestock in the district was twice larger compared to the available grazing land.

Despite the increased number of foreign markets, livestock keepers were incapable of accessing them because of the low quality livestock they have. It was found that more than 90% of the livestock in the district were of indigenous species. Hence most of the livestock keepers in the district depended mainly on the local markets to sell their animals.

Traditionally, livestock were sold in open markets where a number of them were consumed within the district. Some few were transported to other regions. It was reported by key informants that the internal markets do not provide good prices for livestock. All these notwithstanding, it was also revealed that livestock was still one of the major sources of local revenues in the district.

From the interviews it was also shown that, although these species were less productive, they were most preferred by the pastoralists due to their high resistance to disease and resilience to water and food supply stress. In extreme cases, however, livestock deaths do occur (SDC, 2016), as was the case in the dramatic drought of 2008/2009 to 2009/10, which led to the loss of more than 50% of the livestock population (Source: Fieldwork data, 2016).

To summarise, pastoralism in one of the key production sectors, with significant socio-economic values. However, despite the socio-economic contribution of livestock to development of communities and nations, pastoralism is increasingly becoming a tough activity to lead (Ndagala, 1990). Grazing lands are being taken over for other uses, access to water and markets is nowadays very difficult because of the impacts of climate change, and the economic profitability of livestock keeping is critically undermined by the government.

Animals are producing less meat, less milk and are increasingly becoming more and more susceptible to drought and disease. As a result, poverty and resource degradation and conflicts among pastoralists are threatening pastoralism's future which requires more than a positive public relations exercise; it needs to be first placed under scrutiny.

For example, there is a large concentration of pastoral Maasai in the Ngorongoro Conservation Area. Ngorongoro harbours the wildlife based tourist industry and is classified as leading in tourism income generation among other conservation areas in the country. Overall, Tanzania's tourist industry is estimated to be worth between $900m and $1.2bn annually (Oxfam, 2008). Many pastoralists living in these areas support the existence of this industry through their land-use systems. Tanzania has such a high density of wildlife species, both flora and fauna, that it has been classified as one of the 'mega diversity nations' along with the Democratic Republic of Congo, Indonesia, and Brazil.

It is this special biodiversity that draws thousands of tourists to Tanzania each year. Biodiversity in Tanzania is not simply an endowment of nature, it is sustained by the land-use practices and cultural attitudes of the country's people as well as national and international conservation efforts (Oxfam, 2008). However, despite this huge potential, Ngorongoro has one of the highest illiteracy rates in the country.

With the increasing frequency of the effects of climate change and change in the land-use system, the pastoralist Maasai are bound to face increasing levels of poverty and vulnerability. A majority of the households are forced to explore other livelihood support mechanisms to improve household income and ensure sustained support to family needs. Cases of increasing inability of households to mitigate effects of prolonged droughts have been reported within Maasailand. According to a report from Arusha Regional Commissioner's Office, some children starved and others died during the 2013 drought. Some 14 children were treated for malnutrition by the village clinic in Endulen due to the effects of poor diet impacting the pastoralists due to drought.

The absence of an appropriate policy and conceptual framework and a monitoring system to identify and track the true and various contributions of pastoralism to the wellbeing of pastoralists and the economic development of a nation is a major constraint. However, the policy and institutional frameworks to address issues of pastoralism in the East African region are rather differentiated as pointed out by Kateka and Onyango in this book.

Conclusion

Pastoralism is a way of life, an economic system and a tradition among pastoralists. Despite the associated great wealth, pastoralism is not valued by other land users and their governments, and has been mistakenly under-valued by policy makers and governments due to its placement outside the mainstream of national development.

Because of scarcity of land and increased demand for land year after year, conflicts between pastoralists and other land users are common in the East African region. These conflicts have resulted in the deaths of pastoralists and/or their livestock. In Tanzania, recent cases have been reported in Mvomero, Mpanda, Igunga, Kilosa, Meatu, and Malinyi districts where pastoralists have been either shot dead by government agents or other land users. These incidences could have been minimized if land for pastoralism and other land users have been set, demarcated and gazetted.

Other challenges marginalizing the productivity of pastoralism and contribution to the economic development across scales include inadequate stable markets for their products. Meat, milk, hides and skins, and other products, which are a basis for pastoralist wealth and livelihood, and for the economic prosperity of the nation, are undervalued and little has been done to upscale their value. Associated infrastructure to support livestock productivity, through the supply of adequate pasture, water, markets and other requirements for their survival and their livestock are not well established and supported.

Rampant conflicts between pastoralists and other land users are also marginalizing productivity of pastoralism through the loss of livestock and continued mobility of prime animals. Given the prevailing situation in Tanzania today, there should immediately be initiated peace and reconciliation committees in all areas where pastoralists are living.

Generally, a lesson emerging is that the legal protection of pastoralist land rights is not by itself adequate, but that organization and awareness-raising of land uses is important. Pastoralists need to be able to become legally and politically engaged, through civil society organizations and have the confidence and knowledge to be able to stand up for their rights and resolve disputes peacefully. This awareness-raising can be carried out by a range of techniques such as local language broadcasts on rural radios, workshops, community meetings and the training of paralegals from within the pastoral community.

A legal text alone offers a framework, but it is certainly not adequate to change accordingly the behaviour of different actors, to promote pastoralism and sustainable development of a given socio-economic activity and to reduce tensions between different natural resource users. The engagement of vibrant civil society organizations on awareness-raising with regard to local conventions on resource access, establishing forums to encourage social dialogue and the demarcation of livestock corridors are crucial elements that fall outside the scope of this review.

Substantive legislation, that is to say the rules of rights which the courts are called upon to apply, needs to be backed up by solid procedural law or guidelines within which land conflicts or land needs should be accommodated. Also, in addressing disparity and wrong perceptions of pastoralism by different actors, it is necessary to identify sustainable options that enable pastoralists to invest in livestock keeping sustainably while improving their wellbeing and the country's economy at large.

If the policy makers and leaders are to support and recognize as well as appreciate the contribution of pastoralism in the national economies and livelihood of the pastoralists, the pastoral communities will experience a positive change, hence also sustainable development. Informed by the complexity and heterogeneity of pastoral communities, and land-use planning and livelihood strategies, investments and policies need to be matched to accommodate this diversity.

Moreover, detailed research and experience sharing are required to improve our understanding of the conditions under which the economics of pastoralism can be appreciated so as to enhance the economic development of a country, in general, as well as the wellbeing of pastoralists on the ground.

Recommendations

In the East African region, the 1990s have been a busy period of reform of land legislation, democratization, decentralization, and reform of the legislative environment governing pastoralism. There are a number of provisional lessons emerging from this experience that should be shared. It has been learnt, for example, that legislation alone cannot protect livestock mobility, decentralization can restrict as well as expand pastoralist control over strategic natural resources; and that the most effective and adapted natural resource legislation should be rooted and capture the social realities of the entire populations. These lessons provide a benchmark for further recommendations on the improvement of pastoral systems in East Africa as a strategy towards the economic viability of the sub-sector. These may include:

- An optimized use of the rangeland areas for pastoralism in East Africa, which necessitates careful regional land use planning, taking into account the involvement of natives in a given society.
- The development of plans for pastoral livelihoods should involve a mix of interventions. The complementarities between different forms of land-use practices need to be taken into account accordingly for mutual benefit. Policy and investment should also support a diverse range of complementary land uses, as opposed to single or the other land-use option.
- The concept of heterogeneity and homogencity can help planners and decision makers think of the types of investments and supporting policies needed to improve livelihood and enhance national economies. Pastoralism remains an agro-ecosystem that requires system based and holistic planning. Though the complementarities between the economics of pastoralism and the wellbeing of pastoralists are not guaranteed, they require special attention and investments and policies supporting these overlapping land issues.
- The future of the rangelands will in the end largely depend on the development paradigms the society decides to follow for their existence and sustainability. Pastoralism and pastoral lifestyles are unique and tailored to inhabit and use the rangelands efficiently through traditional knowledge systems. The resilient and adaptive knowledge systems of the pastoral people should be incorporated in the national policies and strategies through involvement of the community in planning land utilization, that is to say planning should be bottom up and not top down.

References

Behnke, R. and Muthami, D. (2011). The contribution of livestock to the Kenyan economy. IGAD LPI Working Paper 03-11. Addis Ababa, IGAD Livestock Policy Initiative.

Benjaminsen, T.A., Maganga, F.P. and Abdallah, J.M. (2009). "The Kilosa killings: Political ecology of farmer-herder conflict in Tanzania." *Development and Change*, 40: 423-445.

Blench, R.M. (1998). "Le West African shorthorn au Nigeria." In: *Des taurins et des hommes: Cameroun, Nigeria*, edited by C. Seignobos and E. Thys. Paris, IEMVT, Maisons Alfort; pp. 249-292.

Boyd, C., Blench, R., Bourn, D., Drake, L. and Stevenson, P. (1999). Reconciling interests among wildlife, livestock and people in Eastern Africa: A Sustainable Livelihoods Approach. ODI Natural Resource Perspective, Number 45, June.

Brooks, N. (2006). Climate change, drought and pastoralism in the Sahel. Discussion note for the World Initiative on Sustainable Pastoralism, IUCN.

Centre for International Earth Science Information Network (CIESIN), Columbia University, and International Center for Research on Tropical Agriculture (CIAT) (2004). Gridded population of the world (GPW), version 3 (beta). Palisades. New York: CIESIN, Columbia University. http://beta.sedac.ciesin.columbia.edu/gpw.

Cooper, K. (1998). *Synthesizing research: A guide for literature reviews.* Third ed. Thousand Oaks: Sage.

De Jode, H. (2010). Modern and mobile: The future of livestock production in Africa's dry lands. International Institute for Environment and Development (IIED) and SOS Sahel International, UK.

De Wit. P. (2004). Land and property in the Sudan. Land Tenure Consultant, FAO.

Delgado, C., Rosegrant, M., Steinfeld, H., Ehui, S. and Courbois, C. (1999). "Livestock to 2020: The next food revolution". 2020 Vision Initiative Food, Agriculture and the Environment, Discussion Paper # 28. Washington D.C.: International Food Policy Research Institute.

Desta, S. and Coppock, D.L. (2005). Linking pastoralists and exporters in a livestock marketing chain: Recent experiences from Ethiopia. Research Brief 05-06-PARIMA. Global Livestock Collaborative Research Support Program. NDP Emergencies Unit for Ethiopia 'Cessation of livestock exports severely affects the pastoralist economy of Somali Region, Mission of 31 March to 7 April 1998.

Dyer, N. (2008). Securing pastoralism in East and West Africa: protecting and promoting livestock mobility: Review of the legislative and institutional environment governing livestock mobility in East and West Africa.

Engida, E., Guthiga, P. and Karugia, J. (2015). The role of livestock in the Tanzanian economy: Policy analysis using a Dynamic Computable General Equilibrium Model for Tanzania. International Conference of Agricultural Economists.

Evans-Pritchard, E. E. (1940). *The Nuer.* Oxford: The University Press.

FAO (2009). Grasslands: Enabling their potential to contribute to greenhouse gas mitigation. Submission by the UN Food and Agriculture Organization to the Intergovernmental Panel on Climate Change. Rome: FAO.

FAO (2003). *The state of food insecurity in the World.* Rome: FAO.

Food Security and Nutrition Analysis Unit (FSNAU) (2011). Food security and nutrition analysis: Post Deyr 2010/11. Technical Series Report No. VI. 36. Nairobi: FSNAU.

Greiner, C., Alvarez, M. and Beker, M. (2013). "From cattle to corn: Attributes of emerging farming systems of former pastoral nomads in East Pokot, Kenya." *Society and Natural Resources,* 26: 1478-1490.

Hatfield, R. and Davies, J. (2006). *Global review of the economics of pastoralism.* Nairobi. International Union for Conservation of Nature (IUCN).

Helland, J. (2006). "Pastoral land tenure in Ethiopia." Colloque international 'Les frontières de la question foncière; at the frontier of land issues', Montpellier.

Hesse, C. and Cavanna, S. (2010). Modern and mobile: The future of livestock production in Africa's dry lands. London: International Institute for Environment & Development (IIED) and SOS Sahel International UK.

Hesse, C. and MacGregor, J. (2006). Pastoralism: Dry land's invisible asset? Developing a framework for assessing the value of pastoralism in East Africa. Paper no.42. London: IIED.

Hesse, C. and Thébaud, B. (2006). "Will pastoral legislation disempower pastoralists in the Sahel?" *Indigenous Affairs*, 1/06

Homewood, K.M, Trench, P.C. and Brockington, D. (2012). "Pastoralist livelihoods and wildlife revenues in East Africa: A case for coexistence?" *Pastoralism: Research, Policy and Practice*, 2:19. http://www.pastoralismjournal.com/content/2/1/19.

Integrated Regional Information Network of the United Nations (IRIN) (2006). *Horn of Africa: Pastoralist crisis will not be solved with food aid*. New York: IRIN.

International Food Policy Research Institute (IFPRI) and International Livestock Research Institute (ILRI) (2000). *Property rights, risk, and livestock development in Africa*, edited by B. McCarthy, B. Swallow, M. Kirk and Peter Hazell. Washington, D.C.: IFPRI and Nairobi: ILRI; 433 pp.

International Institute for Environment and Development (IIED) and SOS Sahel International UK (2010). Modern and mobile: The future of livestock production in Africa's dry lands. London: IIED and SOS Sahel International UK.

Khazanov, A.M. (1984). *Nomads and the outside world*. Cambridge: Cambridge University Press.

Knips, V. (2004). Review of the livestock sector in the Horn of Africa (IGAD countries). Livestock Sector Report: Horn of Africa. Rome: FAO.

Lane, C. (1998). *Custodians of the commons: Pastoral land tenure in East and West Africa*. London: Earthscan.

Letara, J., James, M. and Hesse, C. (2006). *Estimating the economic significance of pastoralism: The example of the nyama-choma sector in Tanzania*. Nakuru: Amu Press.

Little, P. (2005). Unofficial trade when states are weak: The case of cross-border commerce in the Horn of Africa. UNUWIDER Research Paper No. 2005/13. Helsinki: United Nations University.

Little, P. and Mahmoud, H.A. (2005). Cross-border cattle trade along the Somalia/Kenya and Ethiopia/Kenya borderlands. Research Brief PARIMA.

McGahey, D., Davies, J., Hagelberg, N., and Ouedraogo, R. (2014). *Pastoralism and the Green Economy – A natural nexus?* Nairobi: IUCN and UNEP; 58p.

Mattee, A.Z. and Shem, M. (2006). Ambivalence and contradiction: A review of the policy environment in Tanzania in relation to pastoralism. IIED Issue Paper No 140, Drylands Programme. London: International Institute for Environment and Development.

Maystadt, J.F. and Ecker, O. (2014). "Extreme weather and civil war: Does drought fuel conflicts in Somalia through livestock shocks?" *American Journal of Agricultural Economics*, 96(4): 1157–1182; doi: 10.1093/ajae/aau010.

Melville, E.K.G. (1994). *A plague of sheep: environmental consequences of the conquest of Mexico.* Cambridge: Cambridge University Press.

Mohammed, A.M. and Faisal, H.R. (2005). "Who controls the water? Rural communities versus local government in North Kordofan State, Sudan," *Haramata*, Vol. 48 (July), IIED, London.

Muhereza, E.F. and Ossiya, S.A. (2004). *Pastoralism in Uganda – People, environment and livestock: Challenges for the PEAP.* Kampala: Uganda National NGO Forum and Civil Society Pastoral Task Force.

National Bureau of Statistics (NBS) (2012). National Sample Census of Agriculture. Volume III: Livestock Sector - National Report.

Ndagala, D.K. (1990). "Pastoralists and the state in Tanzania." *Nomadic Peoples*, 25-27 (IUAES).

Nelson, F. and Ole Makko, S. (2005). "Communities, conservation, and conflict in the Tanzanian Serengeti." In: *Natural resources as community assets: Lessons from two continents,* edited by B. Child and M.W. Lyman. Madison, WI and Washington, D.C.: Sand County Foundation and the Aspen Institute.

Ojalammi, S. (2006). "Contested lands: Land disputes in semi-arid parts of northern Tanzania." PhD Thesis, University of Helsinki, Finland. Available at: http://ethesis.helsinki.fi/julkaisut/mat/maant/vk/ojalammi/conteste.pdf.

Oxfam (2008). Survival of the fittest: Pastoralism and climate change in East Africa. Oxfam Briefing Paper 116.

PINGOs Forum (2016). Impact of Climate Change on the Pastoral Economies of Maasailand. Arusha: PINGOs Forum

Quan, J. and Dyer, N. (2008). Climate change and land tenure: The implications of climate change for land tenure and land policy. Land Tenure Working Paper 2. International Institute for Environment and Development (IIED) and Natural Resources Institute, University of Greenwich.

Simanjiro District Council (SDC) (2016). Socio-economic profile. Simanjiro District Council.

Sulle, E., Lektita, E. and Nelson, F. (2011). From promise to performance? Wildlife Management Areas in Northern Tanzania. Arusha: Tanzania Natural Resource Forum, Ujamaa Community Resource Team and Maliasili Initiatives.

SWAC (2006). Land reform processes in West Africa: A review. Sahel and West African Club Secretariat, Paris.

Swift, J. (1988). *Major issues in pastoral development with special emphasis on selected African countries.* Rome: FAO/UNDP and IDS, University of Sussex.

Trabucco, A. and Zomer, R. (2009). Global aridity index (global-aridity) and global potential evapo-transpiration (global-PET) geospatial database. CGIAR Consortium for Spatial Information.

United Nations Development Programme (UNDP) (2006). Making markets work for the poor. http://www.undp.org/drylands/docs/marketaccess/Making Market Work for Poor.pdf.

URT (2011). Livestock Sector Development Programme. Dar es Salaam: Ministry of Livestock and Fisheries Development.

URT (2010). Livestock Sector Development Strategy. Dar es Salaam: Ministry of Livestock and Fisheries Development.

URT (2006). National Livestock Policy. Dar es Salaam: Ministry of Livestock and Fisheries Development.

URT (2006b). President's speech when inaugurating the Cabinet.

Waters-Bayer, B.A. and Lossau, V.A. (1995). Participatory planning with pastoralists: Some recent experiences. Drylands Programme. London: International Institute for Environment and Development (IIED).

Watkiss, P., Downing, T., Dyszynski, J., Butterfield, R., Devisscher, T., Droogers, P., Pye, S., Ali, B., Harding, B., Tas, A., de Blois, M., Tadege, A., Hunt, A., Taylor, T., Bouma, M. Kovats, S., Maitima, J., Mugatha, S., Kariuki, P., Mariene, L., Worden, J., Western, D., Waruingi, L., Brown, S., Kebede, A., Nicholls, R., Lager, B., Otiende, B., Chambwera, M., Birch, T., Mutimba, S., Sang, S. (2011). *The economics of climate change in the United Republic of Tanzania*. Report to Development Partners Group and the UK Department for International Development. Published January 2011. Available at: http://economics-of-cc-in-tanzania.org/.

Watkiss, P., Downing, T., Dyszynski, J., Butterfield, R., Devisscher, T., Droogers, P., Pye, S., Ali, B., Harding, B., Tas, A., de Blois, M., Tadege, A., Hunt, A., Taylor, T., Bouma, M. Kovats, S., Maitima, J., Mugatha, S., Kariuki, P., Mariene, L., Worden, J., Western, D., Waruingi, L., Brown, S., Kebede, A., Nicholls, R., Lager, B., Otiende, B., Chambwera, M., Birch, T., Mutimba, S., Sang, S. (2009). *The economics of climate change in Kenya*. Final Report to DFID and DANIDA. Led by the Stockholm Environment Institute, Oxford. Available at http://kenya.cceconomics.org/kedo/FINAL-kenyareport-April.pdf.

Wood, A. and Mayer, J. (2001). "Africa's export structure in a comparative perspective." *Cambridge Journal of Economics* 25 (3): 369-94.

CHAPTER 11

A Tale of Three Villages: A Study of Social Stratification in Three Pastoral Communities in Simanjiro District, Northern Tanzania

C.G. Mung'ong'o

Institute of Resource Assessment, University of Dar es Salaam, Tanzania

Introduction

For many years, studies of traditional agricultural production systems emphasized the essential homogeneity of producers. While it was recognized that some differences existed, they were thought, on the whole, not to be significant, but purely a matter of scale (Grandin, *n.d.*).

As noted by Grandin (*n.d.*) and other researchers, earlier research results tended to be reported in terms of average farming households, with average families, average cropping patterns and average yields. In anthropology, to the extent that economic heterogeneity was researched, emphasis was placed on "levelling mechanisms" such as polygamy, polyandry, et cetera. Polygamy has, for example, been discussed as having a levelling function, as more wives would mean more children and consequently the eventual fragmentation of the wealthy producers' assets.

It is worth noting, however, that changes have since been taking place in studies of peasant agriculture, whereby there has been increasing recognition of the extent of wealth differences within communities and their effects on production parameters. As one study noted in northern Nigeria, it was not merely that a few farmers operated on a much larger scale than others, but that there were many rich farmers who had entirely different economic aims from many poorer farmers. The study observed that the tendency of scholars to ignore wealth differences among African peasants had hampered the academia's ability to deepen the understanding of these systems of production.

It is thus that today researchers are paying far more attention to intra-community differences between smallholder farmers. It is recognized that wealthy farmers have differential access to land, labour, animal traction and manure, credit and others. Furthermore, parameters such as savings and investment possibilities make wealthier farmers have different attitudes to risk and innovation than their poorer counterparts.

Despite these developments, recognition of the importance of wealth inequality within pastoralist communities has lagged behind. As noted by Grandin (*n.d.*), much of the household level research in pastoral production systems has continued being done by anthropologists who have paid insufficient attention to the issue of economic inequality in these production systems. The ideology of equality which tends to dominate pastoral societies as well as the apparent similarities in consumption levels of different households, have all contributed to this apparent bias of early anthropologists in studying both peasant and pastoral societies.

This bias has also been fostered by a theoretical tradition which has consistently emphasized the homogeneity of traditional communities. In the so-called non-stratified societies or acephalous societies, there is no concept of social class, power, or hierarchy beyond temporary or limited social statuses. In such societies, every individual has a roughly equal social standing in most situations.

However, even though there are few data available on the distribution of ownership of livestock and other resources within many pastoral communities, the little data that are available indicate that the prevailing pattern of wealth, and consequently income distribution among pastoral societies in eastern Africa, is one of inequality (Konczacki, 1978; Borgerhoff Mulder *et al.*, 2009). Furthermore, Borgerhoff Mulder *et al.* (2010) examined the extent of intergenerational transmission of material wealth, embodied and relational wealth for four pastoral populations from East Africa, West Africa, and southwest Asia and found substantial levels of intergenerational transmission and marked economic inequality.

While there are some indications in the literature of the extent of intra-community wealth differences in pastoral societies, there are few, if any, systematic explorations of the effect of wealth differences on production parameters. Intuitively, we would expect differences in wealth to affect production in many important ways in a pastoral system (Grandin, *n.d.*). Wealth in the form of animal numbers both enables a large family which can be garnered through polygamy, adoption, and the incorporation of poor relations as dependents. Greater wealth also means fewer cash flow problems, and increased possibilities of access to purchased inputs, including drugs and mineral supplements.

Grandin (*n.d.*) has pointed out several points in pastoral systems research at which it would be important to take account of the wealth heterogeneity in the area under study. It would be useful in the early stages as part of the identification of the recommendation domain. Interventions which are useful to large herd owners might not be adaptable by poorer producers. They could, for instance, intensify and even solidify otherwise transient economic inequalities by providing differential access to credit. Secondly, during the verification survey, it is important to be sure that the producers interviewed represent a reasonable cross-section of the population for whom interventions are to be designed.

Hence, Grandin (*n.d.*) argues that the single most important parameter for stratifying within a community is the wealth rank. In conformity with other sceptical commentators elsewhere (cf. IISDnet), Grandin (*n.d.*) warns, however, that "…among pastoralists wealth is a very difficult parameter on which to obtain accurate data. For pure pastoralists, livestock holdings represent a close approximation of wealth." Nevertheless, "(w)hile livestock holdings can be used as a proxy for wealth in a pastoral production system, animal censuses are often difficult, if not impossible, to carry out. Producers are afraid to have their animals counted, due to fear of taxation, other government interferences, or solely on the basis of a cultural taboo on the counting of animals" for fear of the evil eye.

Moreover, there is also a practical caveat to this. Grandin (1983) goes on to explain that "...even if the pastoralist could be persuaded to have his animals counted, the logistic difficulties of such an operation may be overwhelming. Animals may be scattered into different management units, as animals owned ... are not (synonymous with animals kept) ... frequent movements and large distances add to the costs of such censuses. Some of these logistic problems may be overcome by counting animals at watering points, but young and ill animals are likely to be missed. For agro-pastoralists or pastoralists, with significant non-pastoral occupational and investment opportunities, the situation can be far more complex."

Drawing experiences from empirical research done intermittently between 2009 and 2015 in the Simanjiro plains of Maasailand in northern Tanzania, this chapter supports and adds material to the contention that livestock censuses or other objective measurements are not necessary to establish the wealth rank of livestock producers within a pastoral community. The technique of having local informants rank members of their own community according to locally defined wealth criteria can yield similar or even better results at a fraction of the time and expense.

Methodology

Defining the assets

The data which this paper analyses are derived from wealth ranking exercises conducted in three pastoral communities, *i.e.* Namalulu, Kitwai A, and Terrat villages, in Simanjiro District in northern Tanzania. The ranking was performed by teams of selected villagers from each village categorized by age and gender. Lists of assets that were perceived as "wealth" in the villages were, first of all, compiled. Informant ranking was first used by Silverman in 1966 in a sociological study of prestige hierarchies. It has subsequently been used by Grandin (1988) and several other anthropologists to elicit wealth ranks among rural communities.

In all three villages, these included livestock, especially cattle, goats, sheep and donkeys, number of wives and children (with emphasis on male children), a sizeable food reserve to feed the 'boma', involvement in business, especially buying and selling of livestock. Others included employment - that allows one to hire labourers and to send children to school - owning a shop, owning a milling machine, and owning a number of beehives. Ownership of houses for rent, tractors and transport vehicles was an important criterion of wealth in Namalulu and Terrat.

The importance of each of these assets as criteria for assessing wealth or social status of villagers was then determined. It was interesting to note that ownership of land did not feature as an important criterion for assessing the social status of pastoralists. In Kitwai A, for example, land by itself was not an issue. It was assessed in the context of other socio-economic attributes. In the same community the ownership of modern assets such as a permanent house, a motor bike, a mobile phone, *etc.*, was not an issue in Kitwai A, as a majority of the people did not own these items.

The situation was, however, different in Namalulu where agriculture is as important as livestock keeping. Here the amount of land owned and cultivated was an important criterion differentiating people in wealth groups. In Terrat, involvement in mineral brokering and engagement in real estate brokering in Arusha town were other important criteria for assessing wealth among pastoral producers.

Defining the wealth groups

Characterizing wealth groups in the community was, first of all, done by determining the importance of each of the assets mentioned above as criteria for assessing wealth or social status of villagers. This activity was followed by establishing the definition of a "household" among pastoralists. Pastoral systems are marked by the communal sharing of major resources such that the focus on the "household" as a unit of production is often more problematic with pastoralists than with smallholder farmers.

Generally, in order to understand pastoral management practices and strategies, far more attention must be paid to the relationships between individuals and "households" in a social system as discussed below. Hence, the concept of boma or *almarei* was adopted as a unit of analysis of the pastoralist data while the traditional "household" was maintained and applied in the analysis of non-pastoral or agro-pastoral data.

After a protracted discussion on the ranking criteria, the ranking teams then ranked all the names of heads of the boma and/or households. The names were read out aloud from *vitongoji* registers developed by the respective *vitongoji* chairpersons in collaboration with the respective Village Executive Officers (Vitongoji, the sub-village, is the smallest administrative unit in rural Tanzania, headed by a chairman; sing. *Kitongoji*). Placing a particular name in a specific group category was based on consensus of all the participants in the ranking exercise.

Results and Discussion

The socio-physical context

The semi-arid Simanjiro District is situated in Northern Tanzania. Its rainfall pattern, temperatures, topography, soils, vegetation, population density and dynamics, and its administrative structure are variously described in SDC (2005) and Yanda and William (2010). Maasai social organization is adequately described in McCabe *et al.* (2010); Goldman (2003); Hauf (2003) and Homewood and Rodgers (2004); Brockington (2001) and Homewood and Brockington (1999). Suffice to reiterate here that Maasai social organization is based on three interlinking institutions: marriage and family relationships, territory, and the age set.

According to McCabe *et al.* (2010), the smallest social unit is the house or *engaji,* consisting of a wife and her children. A man, his wives and dependents form the next organizational unit, referred to as *engishomi* or *almarei* (the 'gate'). The 'gate' - also popularly known in Kiswahili as the 'boma' - is the basic unit responsible for the management of livestock (Hauff, 2003; McCabe *et al.*, 2010). In the past it was typical for a number of *engishomi* living together in a unit to be referred to as an *engang*. One rather recent development is that there is a trend for an *engang* being defined as a single *engishomi,* but this varies across Maasailand (McCabe *et al.*, 2010). Every Maasai man must be a member of a clan (*olgilata*) which is the basic unit for mutual aid and redistribution of livestock during and after a calamity.

McCabe *et al.* (2010) also mention that the Maasai are organized into territorial sections locally known as *olosho* in which all members have access to the grazing resources. This is the largest political unit for the Maasai, and in the past, such *olosho* could engage each other in war for livestock and/or grazing resources.

The age/grade, age/set system is the other basic unit of Maasai social organization (McCabe *et al.*, 2010). Together with other members of his cohort or age set, each male member of Maasai society goes through three age grades, the *ilayiok* "uncircumcised boys", the *ilmurran* "circumcised young warriors", and the *ilpayiani* "elders". The *ilpayiani* "elders" are themselves distinguished between junior elders, senior elders, and retired elders; each with its own set of norms and responsibilities (Ndagala, 1992 and Talle, 1988 as quoted in Hauff, 2003).

McCabe *et al.* (2010) further inform that during the *ilmurran* warrior grade, members acquire a name that identifies them as an age set and remains with them throughout their lives. Age sets have leaders popularly known as *aleguanan* (plural *aleguanak*), that constitute the political base of the society (*ibid*). As senior elders, the *aleguanak* wield significant power and authority over decisions relating to what is right or wrong with respect to the management of natural resources and articulating with outside entities (*ibid*).

While males move through the age-grades with elaborate rituals, women are merely associated with the system; they are not a part of it (McCabe *et al.*, 2010). For this reason, women are, by definition, "subordinate to men in all aspects of cultural subtlety" (Talle, 1988 as quoted by Hauff, 2003).

Wealth groups and their respective socio-economic characteristics

Table 11.1 illustrates the wealth groups identified and categorized in Kitwai A, Namalulu and Terrat villages with their different characterizations.

Table 11.1: Percent Distribution of Wealth Groups and their Respective Socio-Economic Characteristics in Kitwai A, Namalulu and Terrat Villages, 2009-2015

WEALTH GROUP	SAMPLE VILLAGES
	KITWAI A
IlArkasis kitoo (The most well-off)	Own between 400-500 cattle, 150-200 shoats, and around 10 donkeys. Have between 6-10 wives, and 20-30 children. Able to cultivate between 10-20 acres of maize to supplement food products from livestock.
%	3.3
IlArkasis kinyi (The less well-off)	Just below the *IlArkasis,* having between 100-300 cattle, 50-100 shoats, and about 5 donkeys. Have between 3-5 wives with 10-18 children. Able to cultivate between 3-5 acres of maize.
%	9.5
IlArkasis aikah/ IlOltungani oikash (The average well-off)	Have between 30-100 cattle, 20-30 shoats, and around 2-3 donkeys. Have 1-2 wives, with 5-9 children. Able to cultivate 1-2 acres of land. Involved in the business of buying and selling of livestock, and own milling machines within the village. Have enough food reserves to last the year, but without a surplus. Dependent on family labour.
%	16.1
IlOrkunjai ("The Cattleless")	Group of employed servants within the village, *e.g.* teachers, medical attendants and employees of the tourist tented camp and other itinerant petty business people – also collectively known as *Waswahili.* Also includes traditional beekeepers, mainly comprising of the *Arkiek Dorobo.*
%	29.5
IlMenat (The poorest of the poor)	Have less than 10 cattle. May have a wife and children, but dependent on tending to the IlArkasis cattle. People who have somehow lost their wealth in the above three groups belong to this group. Destitutes (IlAisinak) and many of the aged and the disabled belong here.
%	41.6

Source: Fieldwork Data, March 2009-2015.

NAMALULU	TERRAT
Own between 500-1000 cattle, more than 300 shoats, and around 10 donkeys. Have between 6-8 wives, 20-30 children and more than 10 labourers. Able to cultivate between 50-100 acres of maize. May have a tractor and 1-2 rental houses.	Own between 1000-4000 cattle, more than 500-600 shoats, and around 50 donkeys. Have between 8-10 wives, 40-50 children and more than 10 labourers. Able to cultivate between 50-100 acres of maize. May have a tractor and 1-2 rental houses.
3.4	7.6
Just below the *IlArkasis kitoo*, have between 100-300 cattle, 50-150 shoats, and about 5 donkeys. Have between 3-5 wives with 10-15 children and 4-6 labourers. Able to cultivate between 20-30 acres of maize and beans.	Just below the *IlArkasis kitoo*, have between 100-900 cattle, 100-500 shoats, and about 20-40 donkeys. Have between 5-8 wives with 20-30 children and 5-9 labourers. Able to cultivate between 20-40 acres of maize and beans. Involved in the business of buying and selling of livestock, and own shops and milling machines within the village.
2.2	11.6
Have between 20-100 cattle, 15-20 shoats, and around 2-3 donkeys. Have 1-2 wives, with 5-9 children and 1-2 labourers. Able to cultivate 10-20 acres of land. Involved in the business of buying and selling of livestock, and own shops and milling machines within the village.	Have between 50-90 cattle, 50-90 shoats, and around 10-20 donkeys. Have 3-5 wives, with 10-20 children and 1-5 labourers. Able to cultivate 10-20 acres of land.
9.4	20.5
Employees such as teachers, nursing auxiliaries, petty traders.	Employees such as teachers, nursing auxiliaries, petty traders, mineral brokers.
72.8	34.7
Have only 2-3 cattle, and may have a wife and some children. Own around 10 acres but cannot cultivate them all, loans out some of them to wealthier groups. Beggars and many of the aged and the disabled belong to this group.	Have only 2-4 cattle and some shoats. May have a wife and some children. Own around 10 acres but cannot cultivate them all, loaning out some of them to wealthier groups. Many of the aged and disabled belong to this group.
12.2	25.6

Table 11.1 above raises some very interesting points to social science research. First of all, what is particularly interesting in the fivefold characterization of the wealth groupings in the three communities is the amount of detail that went into their grouping, their characterization and even in their nomenclature. It shows that pastoralists know their communities and each other much better than many of the smallholder farming communities do. They are able to distinguish the three distinct wealth classes, *i.e.* material, embodied and relational, that are important in contemporary social stratification and in future intergenerational transmission of wealth (Borgerhoff Mulder *et al.*, 2009).

The second interesting point is the comparability of the wealth ranking data with data from other more conventional poverty assessments done in Tanzania. If you add up and average the percentages of the *IlArkasis* wealth group and that of *IlMenat* group in Table 10.1 you will realize that they amount to around 5% for the *IlArkasis* and about 50% for the *IlMenat,* respectively. With the distortions brought about by putting the employees and public servants among the *IlOlkunjai* (People-without-cattle) wealth group taken into account, these figures compare very well with figures from conventional poverty assessments such as those by the Research and Analysis Working Group (RAWG) in Tanzania which put the percentages of the poor at between 40% - 55%, depending on the location of the rural area (RAWG, 2005). Note also the similarities of this stratification with the stratification of the US society. According to Beeghley (2004), the super-rich in the US formed 0.9%, the rich were 5%, the middle class ca. 46%, the working class ca. 40% to 45%, and the poor ca. 12%.

Thirdly, it is interesting to note that ownership of land did not feature as an important criterion for assessing the social status of pastoralists. For pure pastoralists, land by itself was not an issue as was the nature of land to provide pastures and watering sources. What was key was one's ability to move animals far and wide during moments of travail such as severe drought. As pointed out by Homewood and Rodgers (2004), natural resources in these communities are managed through common property regimes where access to pastures and water are negotiated and dependent on reciprocal arrangements rather than individual tenure. This is unlike for agro-pastoralists and pure cultivators where land is an issue; as well as access to other productive resources than land that will enhance their ability to cultivate large tracts of such land.

The fourth interesting point is the logic of placing employees and other non-livestock keeping people in lower categories than pastoralists;

categories that also included traditional beekeepers from the almost pariah Arkiek Dorobo. This categorisation can only be explained by the prestige and socio-cultural importance with which cattle are bestowed in the Maasai community.

As noted by Hauff (2003), cattle are a structural backbone of many important stages in the life history of a Maasai. According to Borgerhoff Mulder *et al.* (2010) livestock are the principal form of material wealth among pastoralists, serving as the fundamental form of family capital and identity. They are instrumental in the bonds of personal relationships and necessary for the Maasai way of life. They determine the age at which young people marry, as men need to be independent and own a good herd size before they are able to establish their own matrimonial group (Ndagala, 1992; Hauff, 2003; Homewood and Rodgers, 2004).

As elaborated by Homewood and Rodgers (2004), a Maasai boy gradually accumulates animals held for him in his mother's herd, whether allotted at the time of her marriage, born from allotted cows or acquired later as gifts from her father and other close relatives. According to Hauff (2003) most men start out with a herd of about three cattle. With the estimation that a family needs about 30-50 cattle to barely support itself (*a la* Hillman, 1994), such an accumulation may take about 24 years to achieve (Ndagala, 1992:103); *i.e.* about the time that men enter the *ilmurak* "warrior" grade and are ready to marry. Thus, to a Maasai pastoralist, a man without livestock is not a complete man. As emphasized by Waller (1999), real humans among the Maasai "own cattle, others are just low-grade farming scum."

In terms of class, the Maasai notion of social stratification comes closer to the Weberian conceptualization of social class whereby stratification is not based purely on economic inequalities but on other status and power differentials as well. Max Weber emphasized the importance of political power and social status or prestige in maintaining class distinctions (Weber, 1958). Among the Maasai community, emphasis is placed on ownership and the social meaning of livestock.

Changing patterns in poverty levels

A fourth point from Table 11.1 is that the *IlArkasis* tend to keep large herds of cattle as a strategy to accumulate wealth. However, as noted by Borgerhoff Mulder *et al.* (2010), successful herd management involves social as well as material capital, whereby secure access to water, pasture, and labour requires the establishment and maintenance of supportive social relationships within and beyond the community. Traditionally, when the herds got too big and unmanageable by an individual they split them in small units and loaned them out to either the moderately rich people or the poor to take care of them. In payment the keeper drank the milk and kept some of the heifers born in the new place.

Although in this way the *IlArkasis* continued to be rich, the importance of this system – also known as *mafisa* among the Rangi (Mung'ong'o, 1995) and the Sangu (Borgerhoff Mulder *et al.*, 2010) – lies in its capacity of not only distributing risk between different ecological habitats, it also acted as a wealth sharing system of some sort, playing an indirect in-built mechanism for poverty alleviation among these communities (Mung'ong'o, 1995). The poor benefited and were thus able to positively shift their position on the wealth spectrum to a relatively better wealth category (Borgerhoff Mulder *et al.* (2010).

However, the system has slowly been disintegrating as Maasai traditional institutions have become untenable. For example, with the 2009 drought, incidences of suicide have been on the increase in Maasailand, a phenomenon that was unheard of a few decades ago (Kitwai A Village Chairman, personal communication, 12th March 2010).

The practice in times of severe drought today is that the *IlArkasis* sell a few bulls to hire some young people from the *IlOltungani oikash* and the *IlMenat* groups whose diversification options are limited by the assets they possess to care of and move out with their other animals in case of a severe drought. Unlike earlier times, the latter slowly try to shift their position on the wealth spectrum through selling their labour for wages, or keeping the *IlArkasis* cattle for payment in cattle and earnings from agriculture.

At the time of this study, the market price of cattle per head, depending on the age and health of the animal, ranged between Tsh. 80,000-120,000, while goats and sheep fetched a price of Tsh. 5,000 and 6,000, respectively (Average exchange rate: 1 USD\$=Tsh 1,450). These prices usually dropped drastically in times of prolonged droughts, leading to a spiral of impoverishment for poor pastoralists.

The migrant livestock keepers, on the other hand, were faced with a series of conflicts over water source use, human-wildlife conflicts in protected areas such as Tarangire wildlife dispersal areas (Goldman, 2009), and conflicts between Maasai pastoralists and resident agricultural communities, *e.g.* in Kilosa (Benjaminsen *et al.*, 2009), Kilombero and now Lindi.

Back home, the *IlArkasis* were also diversifying into business, especially through selling of livestock and investing in lucrative mineral and real estate brokering in towns, especially in Arusha in northern Tanzania. According to Lynn (2010) about "…50% of study households participated in gem trading, most breaking even or making a small profit…. While these individuals invest in large-scale village projects…. the purchase of a tractor is the means through which one person's profits have the greatest potential to impact land use change across an entire village as purchased tractors are rented to other villagers to cultivate their own plots…." Resources accrued from such mining and real estate enterprises have also been used to change the pastoralists' way of life, including building of modern houses, buying and owning of motorcycles (*boda boda*), motor vehicles, mobile phones and generators for electrifying their homes in urbanized villages like Terrat and Namalulu.

Crop cultivation and Maasai poverty levels

Currently, more than 80% of Tanzanian Maasai and almost half of the Kenyan Maasai cultivate (Homewood *et al.,* 2001 as quoted in Hauff, 2003). According to a narrative by Waller (1988), few events have historically influenced the current Maasai attitude towards agriculture as much as the period known as *Emutai* that occurred at the end of the 19th Century. This period was characterized by serious outbreaks of alien cattle diseases which devastated much of the cattle herds. After these outbreaks, came an outbreak of smallpox among the human population. The last stage of these disasters was followed by internecine warfare among many of the Maasai groups, as people attempted to rebuild herds by cattle raiding from groups that had managed to prevent wholesale cattle losses due to the diseases.

One important impact of these events was that many people migrated to live with agricultural peoples who were much less affected by the disease outbreaks (McCabe *et al.* 2010). Many men ended up living and marrying within these groups, especially the Gikuyu of Kenya, and the WaArusha and WaMeru of northern Tanzania. Eventually most of these men returned to Maasailand and with them they brought "home" their wives and their agricultural knowledge.

Because of the adoption of agriculture as a livelihood strategy against climate variability/change and the attendant vagaries of nature, at the UN Conference on Desertification in 1977, pastoralists were accused of behaving as if they regarded breeding stock, rather than the land and its vegetation, as their ultimate resource. This was said to be especially so when they were under the environmental stress to which they were increasingly becoming subject. This, according to the Conference, had led to an apparent disregard for the ecology of the plant communities on which their livestock fed.

Contrary to this "blaming the victim" attitude by the UN body, McCabe *et al.* (2010) have recently shown that the motivations for adopting cultivation differed among people of different wealth categories. Besides the influence described earlier that is associated with interactions between pastoralist communities and the non-pastoral Waswahili, it has also been established that pastoralist communities also engaged in agriculture not only to insure their households against food insecurity, they engaged in crop cultivation as a strategy to balance their livestock numbers (Yanda and William, 2010). Data from Namalulu Village in the present study, for example, clearly showed that some of the earnings from agriculture were used to buy more cattle as a strategy for wealth accumulation for the *IlArkasis,* and generally for attempts to move out of poverty for the *IlOltungani oikash* and *IlMenat* wealth groups.

These motivations were as rational and as economic as any you can find elsewhere in the world. According to calculations by Yanda and William (2010), when a pastoralist took up crop cultivation, he already had an insurance of between Tsh. 80,000-120,000 per head of cattle. When he successfully cultivated maize, he had an additional income of between Tsh. 17,000-20,000 per 100 kg bag of maize. This added up to an estimated income of between Tsh. 97,000-140,000 per unit of measure, depending on the number of bags one earned and the number of cattle one had. The common denominator here was, however, the reliability of rainfall in that season and not the ecology of the plant communities on which the livestock fed.

More important, McCabe *et al.* (2003) have also argued that those decisions were part of a larger cultural shift and were also influenced by "...power differentials among Maasai age sets in the context of prevailing government policies". In this respect, McCabe *et al.* (2003) observed that during the last thirty years most of the Maasailand's purely pastoral peoples had diversified their economies. Reasons advanced for this cultural shift in livelihood strategies included the alienation of rangelands due to the expansion of parks and protected areas (Goldman, 2003; Brockington, 2002; and Igoe and Brockington, 1999), changes in land tenure and the privatization of land held as common property (Charnley, 1996; Hillman, 1994), and the penetration of the market economy (Hauff, 2003). Other reasons included loss of livestock due to drought and disease, and the increase in the human population at the expense of the livestock population that had remained steady or slightly declined (Homewood and Rogers, 2004).

Impelled by livestock losses from disease, drought and land alienation, Tanzanian Maasai also began a sizeable urban migration in the mid-1990s. A number of households in the present study, for example, reported their family members as having migrated to other areas in search of alternative economic fortunes. The dominant migrating cohort was the active age 19-30 years.

Diversification of livelihoods, which wealth ranking tries to interrogate, is a common theme in literature. While studies like Little *et al.* (2001) have argued that only the wealthy and poor pastoralists diversified while those in the middle wealth category did not, data from the present study seems to support the thesis by Homewood *et al.* (2009) who have argued that the diversification process affects all segments of pastoral society. The data show, for example, that in agropastoral villages like Terrat and Namalulu, diversification is a process that pervades all the spectrum of wealth groups. They confirm that it is "cyclical, reflecting the life histories of individuals and the developmental cycles of families". The data also confirm that "the process is linear and permanent"; hence supporting the thesis that "the process may eventually define the livelihood strategy of most, if not all, East African pastoral communities" (Homewood *et al.*, 2009).

Conclusion and Policy Implications

This chapter set forth to support and add material to the contention that livestock censuses or other conventional objective measurements are not necessary to establish the wealth rank of livestock producers within a pastoral community. The chapter has shown that the technique of having local informants rank members of their community according to locally defined wealth criteria is versatile enough and can yield similar results cheaply and in a timely fashion.

As stated earlier in this chapter, recognition of the importance of wealth inequality within pastoral communities has lagged behind due to insufficient attention being paid to the issue of economic differentiation and the prevalence of the ideology of equality which has dominated research in pastoral societies as well as the apparent similarities in consumption levels of different households. This chapter has sufficiently shown that such apparent bias is unjustified.

Although, there were few data available on the distribution of livestock ownership within the studied pastoral communities, the chapter agrees with Konczacki's contention that "...the prevailing pattern of wealth, and consequently of income distribution among pastoral societies dependent on animal husbandry, is one of inequality" (Konczacki, 1978). Such inequality is actually recognized by the pastoral communities themselves and determines the characterization of their social relations as evidenced by the data presented in Table 10.1.

And lastly, while wealth differentiation as depicted in this study is an interrogation of the livelihood diversification process in pastoral communities in general, it is also a confirmation of the importance of pastoral systems research to take account of the wealth heterogeneity in the area under study as advocated by Grandin (*n.d.*). To ensure that the 'bomas' or households interviewed represented a reasonable cross-section of the population for whom interventions were being designed, the wealth ranking classifications identified in the present study could be fruitfully used as sampling frames for the household questionnaire surveys.

Also, wealth ranking could be useful in the early stages of a project formulation as part of the identification of the recommendation domain. As argued by Grandin (*n.d.*), clearly "a producer with 400 animals will have different management strategies and possibilities than a producer with only 4 animals". In our case, for example, the *IlAlkasis* pastoralists, particularly those with sufficient labour within a 'boma', had more management options in terms of herd-splitting and lending of animals for risk avoidance and to maintain useful social networks. More important, they also had an added advantage in easy access to important animal inputs necessary for a profitable pastoralist enterprise.

Acknowledgements

Material used in this chapter is an output of the research project "Changing ecosystem functions and human land uses of the savannas of southern Kenya and northern Tanzania", a cooperation between the Institute of Resource Assessment of the University of Dar es Salaam and Michigan State University of USA. It was financially supported by Michigan State University with funds from various funding agencies in the USA.

References

Beeghley, L. (2004). The structure of social stratification in the United States. Boston, MA: Pearson, Allyn and Bacon.

Benjaminsen, T.A., Maganga, F.P. and Abdallah, J.M. (2009). "The Kilosa killings: Political ecology of farmer-herder conflict in Tanzania." *Development and Change*, 40: 423-445.

Borgerhoff Mulder, M., Fazzio, I., Irons, W., McElreath, R.L., Bowles, S., Bell, A., Hertz, T. and Hazzah, L. (2010). Pastoralism and wealth inequality: Revisiting an old question. *Current Anthropology*, 51(1); February.

Borgerhoff Mulder, M., Bowles, S., Hertz, T., Bell, A., Beise, J., Clark, G., Fazzio, I., Gurven, M., Hill, K., Hooper, P.L., Irons, W., Kaplan, H., Leonetti, D., Low, B., Marlowe, F., McElreath, R., Naidu, S., Nolin, D., Piraino, P., Quinlan, R., Schniter, E., Sear, R., Shenk, M., Smith, E.A., von Rueden, C., and Wiessner, P. (2009). Intergenerational wealth transmission and the dynamics of inequality in small-scale societies. *Science*, 326(5953): 682–688; October.

Brockington, D. (2002). *Fortress conservation: The preservation of the Mkomazi Game Reserve, Tanzania*. Oxford: James Currey.

Charnley, S. (1996). "Pastoralism and the demise of communal property in Tanzania." *Cultural Survival Quarterly*, 41-45.

Goldman, M. (2009). "Constructing Connectivity: Conservation Corridors and Conservation Politics in East African Rangelands." *Annals of the Association of American Geographers*, 99(2): 335-359.

Goldman, M. (2003). "Partitioned nature, privileged knowledge: Community based conservation in Tanzania." *Development and Change*, 34(5): 833–862.

Grandin, B.E. (*n.d.*). The importance of wealth effects on pastoral production: A rapid method for wealth ranking. Arid Zones (Eastern and Southern Africa) Programme, ILCA, Kenya.

Grandin, B.E. (1988). *Wealth ranking in smallholder communities: A field manual*. Rigby (UK): Intermediate Technology Publications.

Grandin, B.E. (1983). "Livestock transactions data collection." Pastoral Systems Research in Sub-Saharan Africa Proceedings of the IDRC/ILCA Workshop Held at Ilca, Addis Ababa, Ethiopia.

Hauff, L.E. (2003). The effects of development on the Maasai: A Thesis. Dept. of Sociology, The Honours Program College of St. Benedict/St. John's University.

Hillman, E. (1994). The pauperization of the Maasai in Kenya. *Africa Today*, 41(4):57-65.

Homewood, K.M. and Rodgers, R. (2004). *Maasailand ecology: Pastoralist development and wildlife conservation in Ngorongoro*. Cambridge: CUP.

Homewood, K.M., Thompson, P., Trench, S., Kiruswa, S. and Coast, E. (2009). Community- and state-based natural resource management and local livelihoods in Maasailand. Special issue on Community based Natural Resource Management. *Afriche e Orienti*.

Igoe, J. and Brockington, D. (1999). Pastoral land tenure and community conservation in East African rangelands: A case study from Northeastern Tanzania. Pastoral Land Tenure Series 11. IIED, London.

IISDnet. Participatory research for sustainable livelihoods: A guide for field projects on adaptive strategies.

Konczacki, Z.A. (1978). *The economics of pastoralism: A case study of Sub-Saharan Africa*. London: Frank Cass, U.K.

Little, P., Smith, K., Cellarius, B., Coppock, D. and Barrett, C. (2001). "Avoiding disaster: Diversification and risk management among East African herders." *Development and Change*, 32: 410-433.

Lynn, S. (2010). The pastoral to agro-pastoral transition in Tanzania: Human adaptation in an ecosystem context. SOURCE No 13/2010: Tipping Points in Human Vulnerability. X. Shen, T. E. Downing and M. Hamza. Bonn, Germany, United Nations University Institute of Environment and Human Security, August.

McCabe, J.T., Leslie, P.W. and DeLuca, L. (2010). "Adopting cultivation to remain pastoralists: The diversification of Maasai livelihoods in Northern Tanzania." *Human Ecology*, 38:321–334.

McCabe, J. T. (2003). "Sustainability and livelihood diversification among the Maasai of northern Tanzania." *Human Organization*, 62(2): 100-111.

Mung'ong'o, C.G. (1995). Social Processes and Ecology in the Kondoa Irangi Hills, Central Tanzania. Meddelanden Series B 93, Department of Human Geography, Stockholm University, Sweden.

Ndagala, D.K. (1992). Territory, Pastoralists and Livestock: Resource Control among the Kisongo Maasai. Uppsala (Sweden): Almquist & Wiksell International.

Research and Analysis Working Group (RAWG) (2005). Poverty and human development report, 2005. Dar es Salaam: Mkuki na Nyota Publishers for Research and Analysis Working Group, REPOA, United Republic of Tanzania.

Simanjiro District Council (SDC) (2005). District socio–economic profile. Unpublished MS.

Waller, R.D. (1999). "Pastoral poverty in historical perspective." In: *The poor are not us.* Edited by D. Anderson and V. Broche-Due. Oxford: Currey; pp. 20–49.

Waller, R.D. (1988). "Emutai: Crisis and response in Maasailand 1883–1902." In: *The ecology of survival: Case studies from Northeast African history.* Edited by D. Johnson, and D. Anderson. Lester Crook Academic Publishing/ Westview Press, pp.73–114.

Weber, Max (1958). "Class, status and party", In: *From Max Weber: Essays in sociology.* Edited by H. Gerth and C. Wright Mills. Oxford: Oxford University Press.

Yanda, P.Z. and William, C. (2010). "Livelihoods diversifications and implications on food security and poverty levels in the Maasai plains: The case of Simanjiro District, Northern Tanzania." *African Journal of Environmental Science and Technology,* 4(3), pp. 154-166, March.

PART V

Synthesis and Conclusions

CHAPTER 12

Sustainable Pastoralism in East Africa: A Pipe Dream or Future Reality?

C.G. Mung'ong'o and P.Z. Yanda

Institute of Resource Assessment and Centre for Climate Change Studies, University of Dar es Salaam, Tanzania

Pastoralism is said to have existed in East Africa since sometime between 3000 BC and 2000 BC or more, years ago (Lamprey 1983; Homewood and Rodgers,1991; Spear and Waller 1993; Reid et al., 2005). Today, pastoralist traditional and often sustainable ways of livelihoods are being threatened. Climate patterns are changing in pastoral areas and variability is increasing. Such changes in climate and ecological conditions have also affected land users, as different land uses require appropriate ecological conditions to achieve success. As climate variations ranging from short-term droughts to long-term climate shifts occur in East Africa, the ability of people to maintain their livelihoods in their traditional lands using traditional methods has also been impacted, "...particularly when these fluctuations are layered with other livelihood stresses" (Lynn, 2010).

Agriculture has displaced rangelands; settlements have reduced the seasonal migration of grazing herds; and pastoral economies are undergoing tremendous changes as more grazing land is being gazetted for wildlife protection (Galvin, 2009). In addition, poverty levels in these communities are very high (Thornton et al., 2009). The focus of this book has been to understand the nature and characteristics of pastoralism in the East African region, including the vulnerability and adaptive capacity of pastoral communities in a changing climate and changing policy landscapes.

In Chapters 3 and 4, the book aimed at understanding the East African pastoral production systems under climate change. Our Sudanese colleagues, Abdelaziz Gaiballah and Nancy Ibrahim Abdalla, clearly show that the climate of the drylands is increasingly becoming characterised by scarce and unreliable rainfall, and is becoming more variable and less predictable. High temperatures ensure that much of what does fall is lost in evapo-transpiration, and intense downpours ensure that water runs off in floods. There are also substantial and unpredictable differences in total rainfall between years, within years and between areas in one year, so that neighbouring villages can experience very different crop yields and harvest success.

In Chapter 5, Anthony Egeru and colleagues show that pastoralists in East Africa are highly exposed and are sensitive to climate change and variability. Climate change and variability are said to generally impact on feed crops and grazing systems, including changes in herbage growth, changes in the composition of pastures, changes in herbage quality, and greater incidences of drought that may offset any dry matter yield increases. In general, the chapter establishes that climate is changing with regional climate projections for East Africa, and that over the next 40 years, rainfall and rainfall intensity will increase, temperatures will rise, as successive poor rains become more common with increased incidences of drought. However, the chapter cautions that these projections are not as straightforward as they seem.

Egeru and colleagues also point out that the traditional coping and adaptive strategies have traditionally included expansion of the grazing range. The movement took place over long distances and lasted many days. Some pastoralists who could afford to have used trucks to move their animals. But as shown by Emma Liwenga and Paul Onyango in Chapters 7 and 8, mobility is no longer as easy to undertake as it used to be because of landscape fragmentation. In addition, migrations across national borders have been shown to expose pastoralists to official harassment.

Movements within the same country are also not without their challenges. Mobility within national borders has been shown to be restricted, as is the case with pastoralists' restricted access to protected areas during droughts. At other times, movement through other pastoralist territories has led to violent conflicts over pastures and water.

The authors also show that pastoralists have practiced raiding to take control over grazing areas or to restock the herds lost to drought. It is now, however, becoming apparent that raiding is no longer for restocking herds or controlling pastures. It has become a commercial enterprise and has a long chain of actors stretching all the way to business men in urban centres who place orders for livestock through corrupt officials.

All in all, the drylands have underexploited development potential and the dominant land-use system – *i.e.* pastoralism – has unique adaptive characteristics that, together with the right enabling policies, suggest that climate change can be adapted to and development can be achieved. The study by the Humanitarian Policy Group argued that, of all the natural resource-based land uses in the drylands, pastoralism functioned best within the prevalent context of wide rainfall variability and unpredictability. "... (w)ith the right policies, investment and support, ...pastoralism presents a logical adaptation route in areas of increased climatic variability, and has an important role to play where other livelihoods are likely to fail" (HPG, 2009).

It is in this context that the effects of climate change on the drylands of East Africa pose particular and difficult policy challenges. Both Paul Onyango and Adolphine Kateka in Chapters 8 and 9, respectively, show, for example, that in addition to policy deficiencies, there is also a problem of contradictions in the policies. Pastoralists are thus increasingly becoming vulnerable to drought and other shocks as a result of policies that are at odds with the unique needs of the drylands, and years of poor and inadequate investment in the development of these areas. This affects pastoralist resilience to drought and their adaptive capacity to climate change. Evidence suggests that, alongside improved drought preparedness planning, disaster management structures and risk reduction efforts, "...it would be more effective to enable and strengthen the inherent adaptive capacity of pastoralists, finding ways to encourage their autonomous adaptation, rather than providing adaptation strategies for them" (HPG, 2009).

Despite these constraints, the rangelands of East Africa have continued to contribute significantly to the national economies and to food security for the region. Edmund Mabhuye in Chapter 10 shows how much is contributed by pastoralism. He also shows that pastoralism possesses considerable untapped opportunities that could benefit the region more economically, including, for example:

- Most of East Africa's livestock wealth is kept by pastoralists in the drylands. The bulk of the meat, milk and other livestock products consumed in East Africa come from pastoral areas in the drylands. Exports of livestock and livestock products from East Africa now annually exceed USD $1 billion and more than 90% of this commerce derives from production systems based on mobile pastoralism.
- In Kenya, livestock raised by pastoralists is worth USD $800 million a year. The sector is worth an estimated three-quarter of a billion dollars a year while in Tanzania the sector in the financial year 2014/2015 contributed directly to the economy an amount of Tsh. 6,486,534,052.89 as revenue collections from the livestock sector, leave alone other economic benefits like employment and food security that livestock provided to the people.
- National parks and other protected areas in the region fall predominantly within the drylands (92% of Kenya's protected areas are in pastoral lands). Tourism brings in annual returns of between $900 million and $1.2 billion to Tanzania's economy, and represents 13% of Kenya's GDP and over 9% of Uganda's.

- Grasslands store approximately 34% of the global stock of CO_2 – a service worth $7 per hectare. African grasslands extend to 13m square kilometres and have vast carbon sequestration potential.

Apart from these basically economic benefits, it has also been observed that pastoralism contributes to the health of rangeland ecosystems. It has been noted, for example, that rangeland ecosystems are healthier where mobile pastoralism continues to be practiced effectively. It is argued, for instance, that "...grazing opens up pastures, stimulates vegetation growth, fertilises the soil and enhances its water infiltration capacity as hoof action breaks up the soil crust, aids in seed dispersal to maintain pasture diversity, prevents bush encroachment and enhances the cycling of nutrients through the ecosystem" (HPG, 2009). In many areas in East Africa, the effects of too-little grazing can be clearly seen, as has been the case with bush encroachment that has rendered large areas of Ngorongoro grazing lands unusable as a result of reduced numbers of grazing animals and the cessation of the use of fire in rangeland management.

Despite these considerable contributions, it is noted that drylands of East Africa receive very little investment in terms of development. Most dryland populations lag far behind the rest of the world in terms of human wellbeing and development indicators, and suffer from the poorest economic conditions in the world. Yet, "...with appropriate investment, drylands need not be poor, as has been shown elsewhere in Argentina, Mexico and Israel" (HPG, 2009).

That pastoralism is a rational use of the drylands is a moot point. Pastoralists have responded to and used, or even chosen, to profit from climate variability and change. This "...allows for a vibrant and productive livelihood system in some of the harshest landscapes in the world" (HPG, 2009). Claude Mung'ong'o in Chapter 11 shows, in this respect, how well organized the Maasai social structure is to be able to manage the highly specialised risk spreading strategies to safeguard their herds against drought, floods, disease and social unrest. Strategies like building up herd sizes as insurance against times of hardship, splitting herds across different locations to spread risk, keeping different species and breeds and loaning surplus animals to family and friends "...ensure the rational use of the natural resource base and also develop and strengthen social relations as a form of social capital" (HPG, 2009).

Given its long history of perseverance, pastoralism is definitely here to stay, even if it is only to satisfy a burgeoning protein demand of a rapidly expanding urban population. Even in highly developed societies such as the Scandinavian countries, the reindeer keeping Sami who shares some common elements with the East African pastoralists, "...is today recognized as a viable community after many years of discrimination" (Broadbent, 2010).

The caveat here is the point raised by many commentators elsewhere and emphasized by Onyango and Kateka in Chapters 8 and 9, respectively. This is the ominously unfavourable policy environment for pastoralists. The two chapters have generally shown that policy marginalization of pastoralists is negatively affecting pastoralism in Tanzania more than in any other of the East African countries. We feel, therefore, that there is urgent need for pastoralist voices to be heard and their ideas to be articulated widely within national policies and strategies now more than any other time in our history. As a nation we cannot afford to ignore pastoralism as one of the key national economic activities. Even more important, we cannot address the challenges of pastoralism development in Tanzania without working with other countries in the region, particularly in the area of policy harmonization.

References

Broadbent, N. (2010). *Lapps and labyrinths: Saami prehistory, colonization and cultural resilience.* Smithsonian Institution Scholarly Press; 304 pp. ISBN 978-0-9788460-6-0.

Galvin, K.A. (2009). "Transitions: Pastoralists living with change." *Annual Review of Anthropology,* 38: 185-198.

Homewood, K. and Rodgers, W. (1991). *Maasailand Ecology: Pastoral Development and Wildlife Conservation in Ngorongoro, Tanzania.* Cambridge: Cambridge University Press.

Humanitarian Policy Group (HPG) (2009). Pastoralism and climate change: Enabling adaptive capacity. A synthesis paper. ODI, Regional Pastoral Livelihoods Advocacy Project.

Lamprey, H. F. (1983). "Pastoralism yesterday and today: The overgrazing problem." In: *Ecosystems of the world: Tropical savannas.* Edited by D. W. Goodall. Amsterdam: Elsevier Scientific Publishing Co. Chapter 3.

Reid, R. S., Serneels, S., Nyabenge, M., and Hanson, J. (2005). "The changing face of pastoral systems in grass dominated ecosystems of Eastern Africa." In: *Grasslands of the world.* Rome: FAO.

Spear, T. and Waller, R. (1993). *Being Maasai: Ethnicity and identity in East Africa.* London: James Currey Ltd.

Thornton, P.K., Van de Steeg, J., Notenbaert, A., and Herrero, M. (2009). "The impacts of climate change on livestock and livestock systems in developing countries: A review of what we know and what we need to know." *Agricultural Systems,* 101(3):113-127.

CPSIA information can be obtained
at www.ICGtesting.com
Printed in the USA
LVHW081731220621
690864LV00003B/70